DOT-VNTSC-NASA-07-01
NASA-L-19379

Wake Vortex Acoustic Characteristics and SOCRATES Sensor Performance

Final Report
January 2003 to December 2006

Kevin L. Clark
Hadi S. Wassaf
Frank Y. Wang, Ph.D.
Andrew Gulsrud
John Dunkel

U.S. Department of Transportation
Research and Innovative Technology Administration
John A. Volpe National Transportation Systems Center
Cambridge, MA 02142-1093

July 2007

Prepared for

National Aeronautics and Space Administration
Langley Research Center
Hampton, Virginia 23681-2199

This document is available to the public through the
National Technical Information Service
Springfield, VA 22161

# REPORT DOCUMENTATION PAGE			*Form Approved* *OMB No. 0704-0188*

Public reporting burden for this collection of information is estimated to average 1 hour per response, including the time for reviewing instructions, searching existing data sources, gathering and maintaining the data needed, and completing and reviewing the collection of information. Send comments regarding this burden estimate or any other aspect of this collection of information, including suggestions for reducing this burden, to Washington Headquarters Services, Directorate for Information Operations and Reports, 1215 Jefferson Davis Highway, Suite 1204, Arlington, VA 22202-4302, and to the Office of Management and Budget, Paperwork Reduction Project (0704-0188), Washington, DC 20503.

1. AGENCY USE ONLY (Leave blank)	2. REPORT DATE July 2007	3. REPORT TYPE AND DATES COVERED Final Report January 2003 to December 2006	
4. TITLE AND SUBTITLE Wake Vortex Acoustic Characteristics and SOCRATES Sensor Performance			5. FUNDING NUMBERS Interagency Agreement: IA1-600, Annex 2 Volpe RA: VX63
6. AUTHOR(S) Kevin L. Clark, Hadi S. Wassaf, Frank Y. Wang, Andrew Gulsrud, John Dunkel			
7. PERFORMING ORGANIZATION NAME(S) AND ADDRESS(ES) U.S. Department of Transportation Research and Innovative Technology Administration John A. Volpe National Transportation Systems Center Advanced Surveillance and Communications Division (RTV-4A) Cambridge, MA 02142-1093			8. PERFORMING ORGANIZATION REPORT NUMBER DOT-VNTSC-NASA-07-01
9. SPONSORING/MONITORING AGENCY NAME(S) AND ADDRESS(ES) National Aeronautics and Space Administration Langley Research Center Hampton, VA 23681-2199			10. SPONSORING/MONITORING AGENCY REPORT NUMBER NASA-L-19379
11. SUPPLEMENTARY NOTES			
12a. DISTRIBUTION/AVAILABILITY STATEMENT This document is available to the public through the National Technical Information Service, Springfield, Virginia 22161.			12b. DISTRIBUTION CODE
13. ABSTRACT (Maximum 200 words) This report provides an evaluation of the current state of the SOCRATES sensor and its readiness for use as an operational sensor for active monitoring of aircraft wake turbulence. SOCRATES is a laser opto-acoustic array designed to passively detect and track the sound emissions of aircraft vortices. The evaluation period, January 2003 to December 2006, included two field tests designed to improve understanding of the fundamentals of wake acoustic phenomenology and aspects of the SOCRATES sensor performance.			
14. SUBJECT TERMS wake turbulence, wake vortices, wake acoustics, acoustic detection, aero acoustics, laser microphone, SOCRATES			15. NUMBER OF PAGES 85
			16. PRICE CODE
17. SECURITY CLASSIFICATION OF REPORT Unclassified	18. SECURITY CLASSIFICATION OF THIS PAGE Unclassified	19. SECURITY CLASSIFICATION OF ABSTRACT Unclassified	20. LIMITATION OF ABSTRACT Unlimited

Acknowledgements

The authors thank Wayne Bryant, formerly of the National Aeronautics and Space Administration (NASA) Langley Research Center (LaRC), for his leadership of this effort — particularly for placing emphasis on wake acoustics phenomenology, which is foundational to the Sensor for Optically Characterizing Remote Atmospheric Turbulence Emanating Sound (SOCRATES) concept. This evaluation could not have been completed without the cooperation of the Denver International Airport management and staff. We specifically thank Eric Hall, Operations Manager, and Michael McKee and Annie Christensen of the Noise Abatement Office. We also thank Flight Safety Technologies, Incorporated (FST), and Lockheed Martin Corporation (LMCo) personnel for valuable discussions and technical support.

In the broader technical community, we acknowledge contributions from Charlie Zheng of Kansas State University and Stanislav Gordev of the University of Notre Dame, as well as many helpful discussions with Thomas Holst of the University Corporation for Atmospheric Research (UCAR). Thanks also go to Don Delisi and Lee Piper of NorthWest Research Associates, Inc. (NWRA), Don Bagwell from NASA LaRC, and Scott Morris from Notre Dame for their help in the deployment and understanding of the test instrumentation. Finally, we thank David Burnham of Scientific and Engineering Solutions, Inc. for translating German literature, and Melanie Soares of the Volpe Center for graphical assistance.

DOT / RITA / Volpe Center

Table of Contents

REPORT DOCUMENTATION PAGE	I
ACKNOWLEDGEMENTS	III
TABLE OF CONTENTS	V
LIST OF FIGURES	VII
LIST OF TABLES	IX
LIST OF ACRONYMS AND ABBREVIATIONS	XI
ES. EXECUTIVE SUMMARY	**ES-1**
ES.1 Background: U.S. Wake Mitigation Programs and Wake Sensors	ES-1
ES.1.1 SOCRATES Sensor Development Prior to 2003	ES-1
ES.1.2 FAA-NASA Wake Vortex Program	ES-2
ES.2 DEN03 Test Summary	ES-3
ES.3 DEN05 Test Summary	ES-4
ES.4 Wake Acoustics/SOCRATES Investigation Findings	ES-6
ES.4.1 SOCRATES Status Relative to 2001 Recommendations	ES-6
ES.4.2 SOCRATES Issues for Monitoring Aircraft Operations	ES-7
1. INTRODUCTION	**1-1**
1.1 Aircraft Wake Vortices	1-1
1.1.1 Wake Vortex Fundamentals	1-1
1.1.2 Wake Vortex Sensor Overview	1-2
1.1.3 FAA-NASA Wake Turbulence Research Program	1-3
1.2 Passive Acoustic Wake Sensing	1-4
1.2.1 SOCRATES Project Background	1-4
1.2.2 State of Passive Acoustic Wake Sensing Prior to 2003	1-5
1.2.3 Denver 2003 and 2005 Wake Acoustics Tests	1-6
1.3 Report Overview	1-8
2. DESCRIPTIONS OF DEN03 AND DEN05 TESTS	**2-1**
2.1 DEN03 Test Description	2-1
2.2 DEN05 Test Description	2-7
3. RESULTS OF DENVER TESTS: WAKE ACOUSTIC PHENOMENOLOGY	**3-1**
3.1 Evidence of Vortex-Generated Sound (NASA-DOT Microphone Array)	3-1
3.2 Consistency of Wake Vortex Sound Generation	3-4
3.2.1 NASA-DOT Microphone Array	3-4
3.2.2 DLR Microphone Arrays	3-7
3.2.3 SOCRATES 2003 Prototype	3-8
3.2.4 SOCRATES 2005 Prototype	3-10
3.3 SOCRATES 2003 Prototype Comparison to Other Sensors	3-11
3.3.1 Wake Detection Comparison (Microphone Array and SOCRATES)	3-11
3.3.2 Wake Track Length Comparison (Microphone Array and SOCRATES)	3-12

3.3.3	Wake Track Length Comparison (Pulsed LIDAR and SOCRATES)	3-14
3.3.4	Effect of Runway Crosswind on Detection	3-15
3.4	**Frequency of Most Consistent Vortex Sound**	**3-15**
3.5	**Relationship of Wake Acoustics and Wake Hazard**	**3-18**
3.5.1	Estimating Wake Circulation from Acoustic Power	3-18
3.5.2	Possible Role of Acoustics in Estimating Wake Axial Coherence	3-19
4.	**2005 SOCRATES SENSOR CHARACTERIZATION**	**4-1**
4.1	**Description**	**4-1**
4.1.1	Principle of Operation	4-1
4.1.2	Deployed Configurations	4-2
4.1.3	Hardware Components	4-4
4.2	**Array Beampattern**	**4-5**
4.3	**Effect of RIV Noise**	**4-7**
4.3.1	Internal and Environmental Noise Sources	4-7
4.3.2	Atmospheric Turbulence Data Collection	4-8
4.3.3	RIV Noise and Acoustic Signal Simulation Descriptions	4-9
4.3.4	Simulation Results	4-11
4.3.5	Interpretation of RIV Simulation Results	4-12
4.4	**Near-Field and Atmospheric Effects on Increased Array Size**	**4-13**
4.4.1	Maximum Effective Beam Length	4-14
4.4.2	Aperture Size Limitation by Propagation Effects	4-19
4.4.3	Loss of Coherence Caused by Turbulent Boundary Layer	4-21
5.	**SUMMARIES AND CONCLUSIONS**	**5-1**
5.1	**Summary: Wake Acoustic Phenomenology**	**5-1**
5.2	**Summary: Comparison of Acoustic Sensors to Other Wake Sensors**	**5-1**
5.3	**Summary: Acoustics/SOCRATES Sensor Issues**	**5-2**
5.4	**Conclusions: Acoustics/SOCRATES Suitability as Wake Sensors**	**5-3**
REFERENCES		**R-1**

List of Figures

Figure ES-1   DEN05 SOCRATES Skyward-Looking Configuration (Conceptual) ES-5
Figure ES-2   DEN05 SOCRATES Billboard Configuration (Conceptual) ES-6
Figure 1-1   Example of Wake Vortices for Arriving Aircraft 1-1
Figure 1-2   Nominal Wake Vortex Behavior in IGE Regime for Different Crosswind Speeds 1-2
Figure 1-3   Denver International Airport Diagram 1-7
Figure 2-1   Aerial Photograph of the Denver 2003 Test Site. 2-1
Figure 2-2   DEN03 Test Sensor Availability Timelines 2-2
Figure 2-3   NASA-DOT Microphone Array System Diagram 2-3
Figure 2-4   NASA-DOT Microphone Array Geometry 2-3
Figure 2-5   Microphone Array PSF in Lateral Direction for Two Frequencies 2-4
Figure 2-6   Microphone Array PSF in Longitudinal Direction for Two Frequencies 2-4
Figure 2-7   Aircraft Type Distribution for the Microphone Array Data Set 2-5
Figure 2-8   Meteorological Tower Deployed in the DEN03 Test 2-6
Figure 2-9   CTI Pulsed LIDAR Mounted on Trailer 2-7
Figure 2-10   Four-Beam SOCRATES 2003 Array in the Skyward-Looking Configuration 2-7
Figure 2-11   Aircraft Observed by DEN05 SOCRATES Skyward-Listening Array 2-8
Figure 2-12   Plan View Diagram of the DEN05 Test Site 2-9
Figure 2-13   DEN05 Test Sensor Availability Timelines 2-9
Figure 2-14   Photographs of Two DEN05 SOCRATES Configurations 2-10
Figure 2-15   Photograph of DEN05 Meteorological Tower 2-11
Figure 2-16   Photograph of Microwave Radiometer 2-12
Figure 2-17   Setup for the RIV Noise Experiment 2-12
Figure 2-18   Close-Up of the Hotwire, Cold Wire and Thermocouple Probes 2-12
Figure 3-1   Examples of Acoustic Noise Source Localization Map Behind Aircraft 3-2
Figure 3-2   Comparison of Vortex Trajectories from Microphone Array and Pulsed LIDAR 3-3
Figure 3-3   Horizontal Beamforming Batch Processing for a B-737-300 3-5
Figure 3-4   Wake Vortex Detection Probabilities for NASA-DOT Microphone Array 3-6
Figure 3-5   Layout of DLR-Berlin Microphone Arrays 3-7
Figure 3-6   Wake Acoustics Detection by Aircraft Type and Duration for the DLR Array 3-8
Figure 3-7   Photograph of DEN03 SOCRATES Sub-Array and Calculated Acoustic Beams ... 3-9
Figure 3-8   DEN03 Test Results for Three Sensors 3-9
Figure 3-9   DEN05 SOCRATES Prototype Wake Track Duration Statistics (Ref. 3) 3-11
Figure 3-10   Detection Statistics for SOCRATES 2003 Prototype and Microphone Array 3-12
Figure 3-11   Distribution of Aircraft Having Microphone Array and SOCRATES Tracks 3-13
Figure 3-12   Track Length Excess: Microphone Array over SOCRATES 2003 Prototype 3-13
Figure 3-13   Distribution of Aircraft Having both Pulsed LIDAR and SOCRATES Tracks 3-14
Figure 3-14   Track Length Excess: Pulsed LIDAR over SOCRATES 2003 Prototype 3-15
Figure 3-15   Crosswind Distribution for the DEN03 SOCRATES Dataset 3-16
Figure 3-16   B-737 Wake and Background Spectra for Non-Dimensional Time Intervals 3-17
Figure 3-17   Histogram of the Largest Energy Excess for B-737 and B-757 3-17
Figure 3-18   Comparison of Wake Acoustic Power and LIDAR Circulation for B-737 3-18

Figure 3-19 Comparison of Wake Acoustic Power and LIDAR Circulation for B-757 3-19
Figure 4-1   Illustration of Acoustic Wavefronts Impinging on Optical Beam 4-1
Figure 4-2   DEN05 SOCRATES Skyward-Looking Configuration (Conceptual, Ref. 3) 4-2
Figure 4-3   DEN05 SOCRATES Billboard Configuration (Conceptual, Ref. 3) 4-3
Figure 4-4   Simplified SOCRATES DEN05 Block Diagram 4-4
Figure 4-5   Optical Diagram for a Single Channel of the SOCRATES System 4-5
Figure 4-6   Predicted SOCRATES 100-m Laser Beam Longitudinal Response 4-5
Figure 4-7   Predicted SOCRATES DEN05 Array Vertical Beam Transverse Response 4-6
Figure 4-8   Setup for Collection of Near-Ground Turbulence Data 4-8
Figure 4-9   Methodology for Simulating RIV Noise 4-10
Figure 4-10 Methodology for Simulating Aircraft Wake and RIV Noise SNR 4-11
Figure 4-11 Simulated Eight-Beam Array SNR vs. Wind Speed and EDR 4-12
Figure 4-12 Measured DEN05 Detection Ratio vs. Crosswind (Ref. 3) 4-13
Figure 4-13 Maximum Effective Beam Length vs. Range to a Monopole Source 4-14
Figure 4-14 Beam Response to an Acoustic Point Source vs. Beam Length 4-15
Figure 4-15 Beam Response to a 50-m Acoustic Line Source vs. Beam Length 4-16
Figure 4-16 Laser Beam Response vs. Wake-Beam Orientation 4-17
Figure 4-17 Beam Response to Point and Line Sources vs. Source Range 4-18
Figure 4-18 Representative Wind-Altitude Profile from DEN05 Test 4-20
Figure 4-19 Representative Temperature-Altitude Profile from DEN05 Test 4-20

List of Tables

Table 1-1 SOCRATES Appropriations History 1-4
Table 2-1 DEN03 Sensors/Equipment and Performing Organizations 2-2
Table 2-2 Microphone Array Resolution 2-4
Table 2-3 DEN05 Sensors/Equipment and Performing Organizations 2-8
Table 3-1 SOCRATES Detection Statistics for DEN03 Reported by FST-LMCo (Ref. 3) ... 3-10
Table 3-2 SOCRATES Detection Statistics for DEN05 Reported by FST-LMCo (Ref. 3) ... 3-11
Table 3-3 Duration of a Non-Dimensional Time Interval for B-737 and B-757 3-17
Table 4-1 Predicted DEN05 SOCRATES 8-Beam Sub-Array Transverse 3-dB Resolution 4-7
Table 4-2 Predicted SOCRATES Sub-Array Number of Beams Needed for 20-ft 3-dB Transverse Acoustic Beam Width 4-7
Table 4-3 Billboard Array Recommended Maximum Sub-Array Separation vs. Frequency ... 4-21

List of Acronyms and Abbreviations

AAF	Anti-Aliasing Filter
A/D	Analog to Digital
AFB	Air Force Base
AGL	Above Ground Level
ARMD	Aeronautics Research Mission Directorate
ASOS	Automated Surface Observing System
BOS	Boston Logan International Airport
BPF	Band Pass Filter
CAA	Computational aeroacoustics
C	Centigrade
C/L	Centerline
CLE	Cleveland Hopkins International Airport
CSPR	Closely-Spaced Parallel Runways
CTI	Coherent Technologies, Inc. (now part of Lockheed Martin Corp.)
CW*	Continuous Wave
CW*	Cold Wire
deg	degree(s)
dB	decibel(s)
DEN	Denver International Airport
DEN03	Test conducted at DEN during late summer of 2003
DEN05	Test conducted at DEN during late summer/fall of 2005
DLR	*Deutsches Zentrum für Luft- und Raumfahrt* (German Aerospace Center)
DOT	Department of Transportation
DSP	Digital Signal Processor
e_{cc}	Eccentricity
EDR	Eddy Dissipation Rate
f	Frequency (of acoustic signal)
F	Fahrenheit
FAA	Federal Aviation Administration

---

* CW has two meanings in this document. The correct meaning in every instance should be clear from the context.

FIR	Finite Impulse Response
FST	Flight Safety Technologies, Incorporated
ft	foot or feet
FY	Fiscal Year
hr	hour(s)
HW	Hotwire
Hz	Hertz
in.	inch(es)
IFR	Instrument Flight Rules
IGE	In Ground Effect
JFK	John F. Kennedy International Airport
JPDO	Joint Planning and Development Office
kt	knot(s)
L	Length of SOCRATES laser beam
L_e	Maximum Effective Length (of SOCRATES beam)
LaRC	Langley Research Center
LMCo	Lockheed Martin Corporation
LIDAR	LIght Detection And Ranging
m	meter(s)
mi	statute mile(s)
min	minute(s)
mph	miles per hour
MIT-LL	Massachusetts Institute of Technology Lincoln Laboratory
NAS	National Airspace System
NASA	National Aeronautics and Space Administration
nm	nano meter(s)
NWRA	NorthWest Research Associates, Inc.
OGE	Out of Ground Effect
Pa	Pascal(s)
PSD	Power Spectral Density
PSF	Point Spread Function
R	Range between acoustic source and beam
RITA	Research and Innovative Technology Administration

RIV	Refractive Index Variability
RVR	Runway Visual Range
sec	second(s)
SFO	San Francisco International Airport
SNR	Signal-to-Noise Ratio
SOCRATES	Sensor for Optically Characterizing Remote Atmospheric Turbulence Emanating Sound
SODAR	SOund Detection And Ranging
SOIA	Simultaneous Offset Instrument Approach
SPL	Sound Pressure Level
STL	Lambert – St. Louis International Airport
$T_{MICROPHONE}$	Wake track length for microphone array
T_{LIDAR}	Wake track length for Pulsed LIDAR
$T_{SOCRATES}$	Wake track length for SOCRATES
TC	Thermocouple
TKE	Turbulent Kinetic Energy
T/R	Transmit/Receive
UCAR	University Corporation for Atmospheric Research
U.S.	United States
USB	Universal Serial Bus
V	Volt(s)
VAD	Variable Azimuth Display
VMC	Visual Meteorological Conditions
WS	Work Station
WTRMP	Wake Turbulence Research Management Plan
λ	Acoustic wavelength
μm	micrometer(s) (also known as micron)
μsec	microsecond(s)

ES. Executive Summary

ES.1 Background: U.S. Wake Mitigation Programs and Wake Sensors

ES.1.1 SOCRATES Sensor Development Prior to 2003

The Sensor for Optically Characterizing Remote Atmospheric Turbulence Emanating Sound (SOCRATES) is an opto-acoustic based wake-sensing technology that uses laser beams as linear microphones to passively detect acoustic emanations from wakes. From its initial funding in 1997, SOCRATES has been, almost exclusively, a Congressionally-mandated system development undertaking by a private company, Flight Safety Technologies, Incorporated (FST), with major support from Lockheed Martin Corporation (LMCo). Total federally-appropriated funding designated for SOCRATES has been $34.5M over nine years, first in the FAA's budget (FY1997-FY2000), then NASA's (FY2000-FY2005). Throughout this period, the Volpe Center supported the FAA and NASA in monitoring SOCRATES development.

SOCRATES testing prior to 2003 was conducted at John F. Kennedy International Airport (JFK) and Langley Air Force Base (AFB). The wake vortex science community in 2001 felt that data from the SOCRATES prototype sensor then in use were difficult to interpret, because there was insufficient understanding of both the basic scientific phenomenon of wake acoustic generation as well as the SOCRATES sensor's self-noise and its response to a sound source. There were disagreements in the research community regarding the spectral characteristics of wake vortex sound. Furthermore, the premise that aircraft vortices emit a unique and consistent acoustic signature that could be utilized as the basis of a wake vortex sensor was not widely accepted by the research community.

Because of these concerns, a report (Ref. 1) was issued by the Volpe Center in 2001 with the following recommendations:

Recommendation 1: Any further development of SOCRATES sensor technology and other candidate systems dependent upon the hypothesized wake vortex sound-generation phenomenon should be deferred until such time as there is a strong, well-established phenomenological basis for their further development.

Recommendation 2: If there is any interest by the Government in pursuing phenomenological research, that is, determining whether wake vortices emit unique, consistent acoustic signatures, for possible application to wake vortex detection, the investigation should be initiated and conducted prior to the development of any potential passive acoustic wake vortex sensor.

Recommendation 3: The FAA and NASA should cooperatively develop a set of requirements for an operational wake vortex sensor to support the development and deployment of a wake turbulence system for mitigation of wake turbulence constraints on airport capacity.

At about the same time, an independent development in wake acoustics research was reported by the *Deutsches Zentrum für Luft- und Raumfahrt* (DLR, the German Aerospace Center) while working under the European Union's C-Wake program. A 64-element microphone array was used to study the feasibility of detecting and tracking aircraft vortices via a passive wake acoustics approach. During field tests, two tracks of noise spots were discernable after aircraft flybys, with motion and spacing consistent with wake vortex physics. The DLR study utilized only a small sample of flights and the array gain was low. However, the DLR experience would be influential in the direction of U.S. wake acoustics research.

ES.1.2 FAA-NASA Wake Vortex Program

Separate from the SOCRATES sensor development, the FAA and NASA developed a cooperative Wake Turbulence Research Management Plan (WTRMP, Ref. 2) that defined an approach to finding solutions which meet growing needs for capacity increase/delay reduction. The FAA/NASA WTRMP, initially released in 2003, envisioned a three-phase incremental approach to evaluating and deploying wake solutions having high benefit-cost ratios and low developmental risks:

- **Near-term Phase (2003-2008):** Develop/implement procedural changes to current arrival operations to Closely-Spaced Parallel Runways (CSPR) without operational deployment of new/additional weather or wake sensors. Specific activities include:
    - Supporting development of the new Simultaneous Offset Instrument Approach (SOIA) procedure at San Francisco International Airport (SFO);
    - Developing new approach procedures with aircraft weight-class restrictions at Lambert-St. Louis International Airport (STL), Cleveland Hopkins International Airport (CLE), and Boston Logan International Airport (BOS); and
    - Modifying the runway separation rule for CSPR approaches applicable to the entire National Airspace System (NAS) by aircraft type.
- **Mid-term Phase (2006-2012):** Develop/implement weather-sensitive procedural alternatives for both arrivals and departures that may involve operational deployment of additional proven weather sensors. Tools will be developed that factor in cross-wind predictions to allow arrival and/or departure operations to be conducted with reduced wake-based separation minimums.
- **Long-term Phase (2010-2015):** Develop/implement active wake solutions that may involve operational deployment of additional weather and wake sensors. The ultimate long-term solution would be an active wake-based separation tool involving real-time monitoring of aircraft, weather, and wakes.

The WTRMP envisioned that the FAA's efforts would be focused on near-term and mid-term solutions that could be implemented through procedure modifications based on extensive pre-implementation wake and weather data acquisition/analysis, and possibly the operational deployment of proven weather sensors. NASA's emphasis would be on fundamental research to enable mid-term and far-term applications. Any SOCRATES acoustic sensor development judged appropriate by the Government would best fit into the far-term applications area.

Recent changes in NASA's Aeronautics Research Mission Directorate (ARMD) research/program emphasis may require that the WTRMP cooperative arrangement be revised or abandoned. However, the phased development process envisioned by the WTRMP remains a technically and programmatically sound approach to an important issue that has been coordinated with stakeholder groups in the U.S. and with international aviation organizations. It continues to be followed by the FAA.

ES.2 DEN03 Test Summary

Test Scenario — To address concerns with the SOCRATES sensor and wake acoustics in general, and utilizing knowledge gained from the DLR test, the Volpe Center conducted a fundamental wake acoustics measurement campaign in August and September of 2003. The campaign was sponsored by NASA's Langley Research Center (LaRC) and was performed at Denver International Airport (DEN). DEN was selected for several reasons: high traffic level, significant fraction of the fleet large enough to have strong wakes, located approximately 30 mi from the city in a quiet rural environment, ample space available for deploying equipment, and cooperative airport authority and local FAA offices.

Equipment was installed north of the airport, where they were positioned to capture sound from aircraft approaching Runway 16L, approximately 2.7 mi from the threshold. At this site, vortex acoustic emanations were frequently audible by personnel on the ground shortly after flyover.

The DEN03 test primary objective was to utilize a 252-microphone array deployed by the Government to establish the scientific foundation of wake acoustics. Additional sensors included: a Pulsed LIght Detection And Ranging (LIDAR) device and a Continuous Wave (CW) LIDAR, for wake "truth"; a meteorological tower, for weather information; a SOund Detection And Ranging (SODAR) device, for wind information; a sonic anemometer, for characterizing atmospheric turbulence; and a microwave radiometer, for characterizing atmospheric stability.

Wake sensing by passive acoustics is simplest to implement when the sensors are located directly beneath the aircraft course of flight. The signal propagation paths are then primarily vertical, and are only minimally changed by variations in the air column. Acoustic sensor performance is best when wake sound appears to emanate from a line source — i.e., the vortex diameters are much smaller than their distances from the sensor. To achieve these conditions, all acoustic sensors were located approximately 700 ft beneath the glide slope, where wakes were generated in the Out of Ground Effect (OGE) regime. All microphone data processing was done after the test, and included examinations of the effects of different frequency bands (e.g., 0-200 Hz and 200-400 Hz), beamforming methods, and modeling techniques (e.g., wavelets).

Microphone Array Performance — Post-test analysis of microphone data using image processing techniques confirmed that aircraft wake vortices do generate acoustic signatures whose characteristics are consistent with wake behavior derived from flow velocity measured by sensors such as LIDARs — e.g., the distance separating an aircraft wake pair and the lateral transport and sink rates of the vortices. However, durations of Pulsed LIDAR wake tracks were generally significantly longer than those for the microphone array, raising concern that hazardous wakes may not always be detected by acoustic means. (However, truncation of the data collection interval for each flyby, as well as use of a non-optimal processing frequency band, could have artificially shortened the microphone array tracks.)

SOCRATES Performance — The DEN03 test also provided the opportunity to examine improvements made to the SOCRATES sensor relative to the prototypes deployed at JFK and Langley AFB. A four-beam SOCRATES array was deployed, along with a smaller microphone array from DLR under contract to FST. Field tests demonstrated that the SOCRATES array was also able to detect and track the source direction of wake acoustic emissions. However, due to the low resolution of its array response pattern, the SOCRATES prototype was not able to distinguish individual vortices; instead it detected the composite turbulence oval encompassing both vortices. With only a single elevation angle, the wake height above ground could not be found.

For statistical comparison purposes, a common set of 213 flybys was identified when the microphone and SOCRATES arrays were simultaneously collecting data. For this dataset, microphone beam forming results for 200-400 Hz were used, as this band is somewhat similar to the SOCRATES' processing band of 270-360 Hz. Overall, detection probability for the microphone array (86%) was a bit better than that for the SOCRATES prototype (70%) for the common flybys. (The SOCRATES contractor team has reported that, for all 882 flybys of the 2003 prototype, a detection probability of 81% was achieved.) Of the 213 common flybys, 83 had track length available from both microphone array and SOCRATES sensor. Average microphone array tracks were longer by 8 to 29 sec, and the difference was almost monotonically increasing with aircraft specified maximum landing weight.

Frequency of Wake Sound — The most consistent frequency ranges of wake acoustic signals and ambient background noise were examined using the microphone array data. The wake signal frequency range most consistently above the ambient background noise was almost entirely less than 200 Hz, with a peak in the 25-150 Hz range. Moreover, the predominance of lower frequencies was more pronounced for larger aircraft.

This finding has significance for the design of potential future wake acoustic sensors. It suggests that — for applications where a larger distance between the wake and the sensor is needed (e.g., at the stabilized approach point 1,200 to 1,500 ft Above Ground Level [AGL], which may be a "wake hot spot" at some airports) — the SOCRATES processing frequency band (270-360 Hz) could limit wake detection performance.

ES.3 DEN05 Test Summary

Test Scenario — The DEN05 test had two main purposes: (1) developmental testing by FST and LMCo of a new prototype SOCRATES sensor; and (2) measurements by the Government for the purpose of characterizing the impact of the atmosphere on SOCRATES performance. In the late summer of 2005, a 16-beam SOCRATES array, arranged in two configurations, was tested at DEN, again to capture data from aircraft approaching runway 16L. A Pulsed LIDAR served as the wake "truth" sensor, and a suite of meteorological sensors was also deployed.

The Volpe Center's focus was characterization of a specific type of atmospheric fluctuation, Refractive Index Variability (RIV), which may significantly impact SOCRATES performance. RIV characterization testing was done using hot wire/cold wire sensors, three sonic anemometers, and three fine-wire thermocouples. A second Volpe Center activity was investigating the impact of atmospheric propagation, which could be an issue if the SOCRATES sensor were to be operated in a mode that listens for a noise source from a standoff-location rather than from directly beneath the flight path.

SOCRATES Skyward-Looking Configuration — The laser beams were deployed in two eight-beam sub-arrays, arranged horizontally (similar to the DEN03 array), one on each side of, and oriented parallel to, the runway centerline (Figure ES-1, Ref. 3). Having elevation angle measurements from two sub-arrays enabled vertical and lateral (i.e., orthogonal to the extended centerline) detection and tracking of wake locations.

The DEN05 sensor, with more beams and improved data processing, provided better angular resolution then the DEN03 prototype. However, like the DEN03 prototype, the DEN05 array detected/tracked the turbulence oval, and was not able to resolve individual vortices.

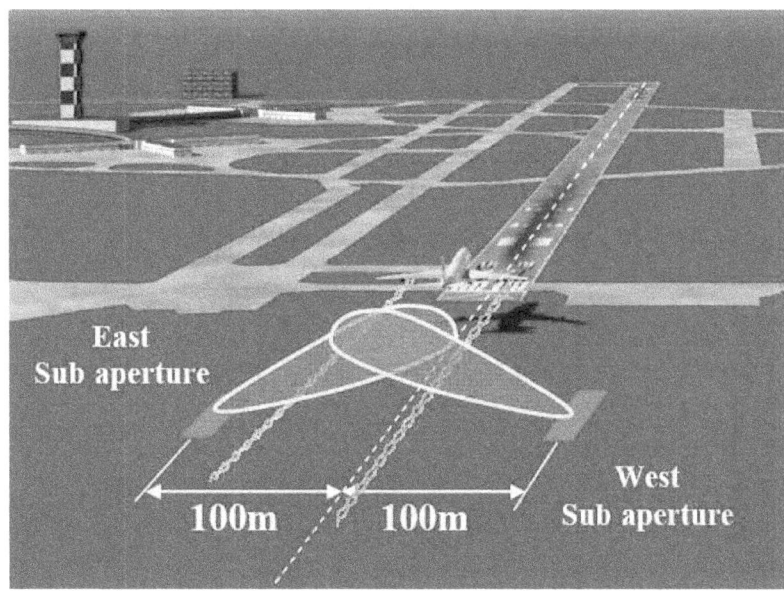

Figure ES-1 DEN05 SOCRATES Skyward-Looking Configuration (Conceptual)
Source: FST-LMCo.

Wake detection statistics from the DEN05 Skyward-Looking array configuration were compiled by the FST-LMCo team. For a dataset of 1,987 flybys (more than double the size of DEN03 dataset), the overall detection rate was reported as 86%. These results have not been independently verified by researchers outside of the SOCRATES contractor team, but improved statistics relative to the DEN03 would be anticipated. What is unexpected is the modest increase (over the 81% for DEN03) in the detection rate, since four times as many beams were available (16 versus 4) and more advanced processing algorithms were used. Relative to Pulsed LIDAR wake tracks, SOCRATES tracks were significantly shorter in duration.

The DEN05 test was conducted under generally favorable circumstances for acoustic wake detection. In addition to the quiet location, clear/dry low-wind Visual Meteorological Conditions (VMC) prevailed almost uniformly. Nevertheless, detection performance was found to be quite sensitive to: (a) crosswind speed (probability dropped to 10% for 10 kt crosswind), (b) solar heating (probability dropped to 60% at mid-day), and (c) temperature (sharp decrease in probability above 80 deg F). The sensitivity to heating and temperature are believed to be related to elevated RIV noise. No data were collected for rain, snow or fog conditions.

SOCRATES Billboard Array Configuration — The laser beams were vertically stacked on concrete towers (Figure ES-2) for standoff listening to wakes. The Billboard array concept was first tested in the DEN05 experiment, where the focal point was approximately 1,000 m (3,280 ft) from the sensor and 14,046 ft from the runway threshold.

According to FST-LMCo, their analyses of the Billboard array data were inconclusive. The Contractor team has stated that there is circumstantial evidence that wakes from one flyby were detected and tracked. No results have been presented to the Government.

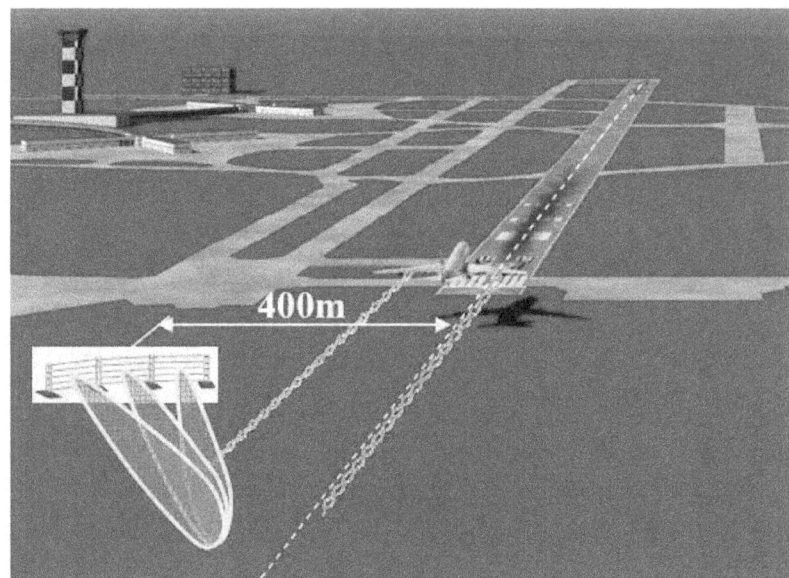

Figure ES-2 DEN05 SOCRATES Billboard Configuration (Conceptual)
Source: FST-LMCo

ES.4 Wake Acoustics/SOCRATES Investigation Findings

ES.4.1 SOCRATES Status Relative to 2001 Recommendations

SOCRATES developmental status is assessed relative to the recommendations of the Volpe Center 2001 report (Ref. 1).

Recommendation 1: Any further development of SOCRATES sensor technology and other candidate systems dependent upon the hypothesized wake vortex sound-generation phenomenon should be deferred until such time as there is a strong, well-established phenomenological basis for their further development.

The phenomenological basis for wake detection by passive acoustic means — whether an array of microphones, an array of SOCRATES laser beams, or another sensor — has been partially established by the DEN03 and DEN05 tests. At the macro level, it has been demonstrated that the wakes shed by many aircraft on final approach emit energy that can be detected by acoustic means. Image processing techniques applied to microphone data revealed that the sound emanated from the two wake vortices and not from the aircraft.

However, the mechanism(s) by which wake sound is generated has not been established. It is conjectured that the low-frequency sound energy is caused by unsteady vorticity in the vortex cores, and that the higher-frequency acoustic energy is due to turbulence in the region surrounding each core. Lack of an understanding of wake sound generating mechanisms limits designers' abilities to predict sound power levels and frequency spectra and their evolution over a wake lifetime. This information is of significant, and perhaps critical, value in the design and certification of a wake detection system, and in predicting its performance sensitivities in environments where the sensor has not been tested.

Recommendation 2: If there is any interest by the Government in pursuing phenomenological research, that is, determining whether wake vortices emit unique, consistent acoustic signatures, for possible application to wake vortex detection, the investigation should be initiated and conducted prior to the development of any potential passive acoustic wake vortex sensor.

Since this recommendation was made in 2001, the Government has pursued parallel efforts, investigating wake acoustic phenomenology while developing the SOCRATES sensor. The phenomenology investigation has been carried out by an array of Government, industry and academic institutions with a combination of NASA program funding and funds earmarked for SOCRATES. The SOCRATES sensor development has been pursued by the FST-LMCo team under Congressionally-earmarked funds.

Recommendation 3: The FAA and NASA should cooperatively develop a set of requirements for an operational wake vortex sensor to support the development and deployment of a wake turbulence system for mitigation of wake turbulence constraints on airport capacity.

The FAA and NASA have not developed requirements for an operational wake sensor. Such a sensor would be part of far-term solution as defined by the WTRMP. However, they have developed operational concepts for near- and mid-term solutions. The applicable weather conditions which support these concepts include moderate to strong crosswinds (5-10 kt) needed to advect wakes from arrival or departure corridors, and are the kind of conditions in which the SOCRATES sensor has poor performance. The weather conditions also include moderate-to-strong background turbulence required to decay vortices more quickly than under calm conditions. This backgrounds turbulence is often correlated with the solar heating conditions in which the SOCRATES sensor also performs poorly.

The FAA-NASA Wake Vortex Program is currently pursuing near-term and mid-term wake solutions which do not involve a real-time operational wake sensor. This development/investment strategy was chosen to take advantage of recently gained knowledge of wake behavior without incurring the cost, technical risks and schedule/regulatory issues associated with a sensor development effort. The use of wake sensors to collect wake data under the applicable weather conditions is considered a preliminary opportunity for evaluation of the sensors as potential components of an operational system for a far-term solution. With the current level of knowledge of wake acoustic phenomenology, no passive acoustic sensor is sufficiently mature for use as a research sensor. For example, the selection/treatment of beam length, number of beams in an array, performance sensitivity to atmospheric conditions, RIV noise mitigation and appropriate model for the spatial/temporal structure of the wake acoustic signal all have significant unknowns/uncertainties. Until basic performance characteristics of the SOCRATES sensor can be established, an occurrence which is estimated to be several years away, an opto-acoustics based wake sensor cannot considered to be a viable option for operational use.

ES.4.2 SOCRATES Issues for Monitoring Aircraft Operations

Limited Resolution Capability — The SOCRATES prototype sensors deployed in the DEN03 and DEN05 tests were unable to separately distinguish the two vortices generated by an arriving aircraft. Instead they detected the oval encompassing the two wakes. Most other wake sensors (e.g., Pulsed LIDAR, SODAR, anemometer windline) can separately distinguish the vortices.

Inability to Determine Wake Strength (Circulation) — A relationship between wake strength (circulation) and wake sound has not been established. In contrast, other wake sensors provide some indication of circulation, with the wake SODAR and CW LIDAR being best. Thus, when an acoustic wake sensor such as SOCRATES loses track of a wake, it is not possible to determine whether or not the wake poses a hazard to other aircraft. Moreover, in several instances, regional jets had stronger wake acoustic signals than Airbus A-318, A-319, A-320 and A-321 aircraft models, which (based on weight) are considered to have more hazardous wakes.

Acoustic Frequency Utilized — To avoid low-frequency internal system noise, the DEN03 and DEN05 SOCRATES prototypes generally utilized the 270-360 Hz band for processing. However, for larger aircraft, wake acoustic signal energy primarily lies in the 0-200 Hz band that can be used by a microphone array. In the quiet Denver environment this mismatch between signal and processing bands may not have had a significant impact on SOCRATES detection performance. In other measurement scenarios — e.g., involving louder background noise or greater distance between the wake and sensor — it is likely that a microphone array would have significantly better performance.

Need for Acoustically Quiet Environment — The rural area surrounding DEN is among the quietest for a major U.S. airport (e.g., much of the time a human observer could hear arrival aircraft wakes). Since this low ambient noise level will not prevail near many busy urban/suburban airports, it is likely that acoustic sensor performance (i.e., wake detection probability and track duration) at such airports would be significantly poorer than it was for the DEN tests.

Sensitivity to Wind and Heating Conditions — Wake acoustic/SOCRATES testing to date has demonstrated detection probabilities in the 80% range during favorable (low-wind, cool) meteorological conditions. However, during the DEN05 tests, SOCRATES detection performance degraded significantly for less-favorable conditions:

- Detection probability began to drop sharply with a crosswind of 5 kt, declining to approximately 10% in a 10 kt crosswind. It is likely that the wind transported the wake at a faster rate than the SOCRATES sensor could detect/track it
- In the DEN05 test, detection probability decreases to approximately 60% were observed in the presence of solar heating and/or elevated temperature, which are similar/correlated effects. The apparent mechanism was an associated elevation of the SOCRATES sensor noise.

These sensitivities might be acceptable in a research context, but would have to be addressed for an operational sensor.

Available Real Estate Under Approach Path — In the Skyward-Looking configuration, the DEN05 SOCRATES sensor required two patches of land near the extended centerline of the runway being monitored, approximately 2.7 mi from the threshold. Each patch measured approximately 330 ft by 10 ft. Similarly, the 252-element microphone array required a 100 ft by 400 ft area under the glide slope. In urban settings, patches at these specific locations may be difficult to acquire. Moreover, for runways having over-water approach paths, installation of an acoustic/SOCRATES wake detection sensor might not be possible.

May Only be Able to Track OGE Wakes — Based on their measurement technique, passive wake acoustic sensors, including SOCRATES, appear to be effective only for detecting wakes that are generated in OGE conditions several hundred feet above the sensor on the ground. The SOCRATES "Billboard" configuration, intended to detect and track wakes near the ground where they may pose the greatest hazard to aircraft, has not demonstrated this capability.

1. Introduction

1.1 Aircraft Wake Vortices

1.1.1 Wake Vortex Fundamentals

Wake vortices, counter-rotating flow structures trailing behind an aircraft in flight (Figure 1-1), are an unavoidable consequence of lift generation. Wakes created by aircraft at least several wingspans above the ground — termed the Out of Ground Effect (OGE) regime — normally descend until either:

- They dissipate, while transporting laterally with the crosswind, without ever descending to within half a wingspan of the ground. Typically, this occurs when the wake is generated many wingspans above the ground.
- They reach an altitude of approximately a wingspan distance from the surface, where the In-Ground Effect (IGE) regime begins and more complex rules govern their behavior.

Figure 1-1 Example of Wake Vortices for Arriving Aircraft

In the IGE regime (Figure 1-2), in the absence of a crosswind, the two wake vortices separate and transport laterally in opposite directions away from the flight path at a ground speed U_g of 3 to 5 kt. In the presence of a crosswind U_{cw} the downwind vortex transport speed will be increased and the upwind vortex transport speed will be reduced. With sufficient crosswind, the upwind vortex may stall or reverse direction. Wakes in the IGE regime generally decay signifycantly more rapidly than those in the OGE regime.

The strength of an individual wake vortex (technically, the angular moment-like quantity termed the circulation) can be significant for large-sized aircraft, and must be permitted to dissipate or transport (laterally or vertically) away from the flight paths of nearby aircraft in order to ensure their safety. Aircraft separation standards (Ref. 4) implemented by the Federal Aviation Administration (FAA) thus include buffers (additional distances) to ensure wake avoidance. However, if the wake vortex imposed separations between a generating aircraft and other aircraft are set too conservatively, then runway and airport capacity will be adversely impacted.

Wake vortex separation standards currently implemented by the FAA, although proven to be operationally safe when proper flight procedures are followed, are considered to be overly

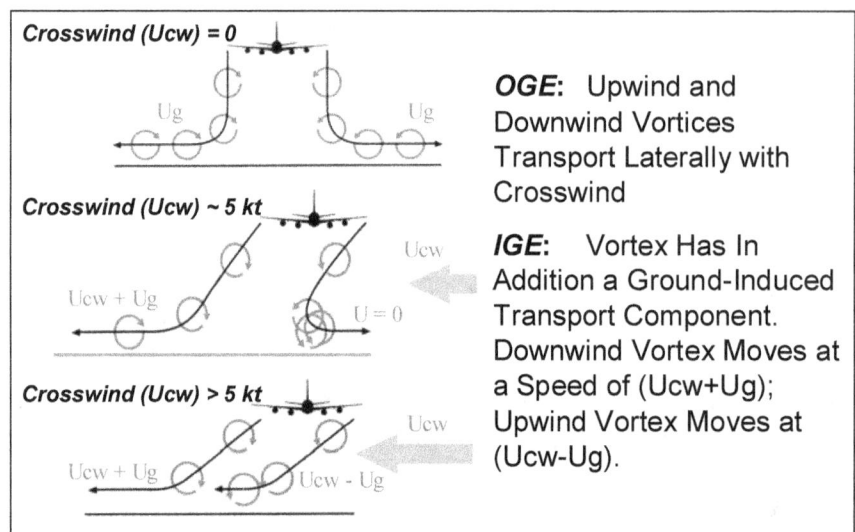

Figure 1-2 Nominal Wake Vortex Behavior in IGE Regime for Different Crosswind Speeds

conservative for certain meteorological conditions and operational procedures. Consequently, opportunities exist to improve airport capacity while maintaining or even enhancing the current level of operational safety. To support the identification and proper documentation of these opportunities, a suite of sensors capable of measuring the vortex position and strength, as well as wind, temperature and turbulence profiles, are needed.

1.1.2 Wake Vortex Sensor Overview

A number of available sensors — other than those relying on passive acoustics, the subject of this report — can provide information concerning wake vortex characteristics:

- Pulsed LIght Detection And Ranging (LIDAR) devices can be used to track wake vortices at relatively long distances (kilometers) and provide estimates of vortex lateral and vertical positions as well as strength. Their primary limitation is that they cannot penetrate through thick clouds and have degraded coverage in rain and fog. Operating wavelengths are 1.5 to 2 µm, which are considered eye-safe.
- Continuous Wave (CW) LIDARs can be used to track wake vortices at distances up to approximately 200 m, a region where wake vortices transition from OGE to IGE. They have better spatial resolution than Pulsed LIDARs, and provide information concerning the vortex structure and strength. However, they have the same visibility limitations as Pulsed LIDARs. In addition, CW LIDARS typically use high power gas lasers, so that eye-safety issues must be considered.
- Anemometer windlines are used to detect and track wake vortices in the IGE regime, and provide reliable lateral position information. Their principal limitation is that they can only measure wakes near the ground. Moreover, even for IGE wakes, the estimates of strength and height are not accurate. The exception is a windline anemometer array using tall poles; then the cores of IGE vortices can at times traverse through the anemometers, thereby providing structure and strength information. The anemometers can be either propeller or sonic types. They are generally not susceptible to failure in adverse weather conditions except for ice accumulation (propeller type) and heavy rain (sonic type).

- SOund Detection And Ranging (SODAR) devices are essentially the acoustic version of the Pulsed LIDARs that sense the vertical wind field above the instrument. Multiple units can be deployed as an array, in a manner similar to windlines, to provide estimates of vortex strength and lateral position. However, SODARs are limited in vertical range to a few hundred feet and are subject to atmospheric effects that significantly affect sound propagation. Airport noise and noise generated by wind and rain can also degrade performance.

All of the above sensors have been used as research instruments (e.g., Ref. 5), and (with the exception of windlines) are available commercially. All of the sensors fundamentally measure the wind field using various principles and detect the velocity signature of wake vortices. As such, the wake data collection also contains meteorological information. At present, their practicality as operational sensors for real-time use in controlling air traffic is problematical for several reasons, including performance limitations, lack of maturity, siting constraints, complexity of operation, insufficient reliability, and lack of standards/requirements. Use of wake sensors for changing aircraft separation minimum distances in real-time has not been attempted by the FAA.

1.1.3 FAA-NASA Wake Turbulence Research Program

Currently, the FAA and the National Aeronautics and Space Administration (NASA) are conducting a joint Wake Turbulence Research Program (Ref. 2). The overall research effort distinguishes among near-term, mid-term, and far-term activities.

Near-term efforts improve airport capacity through procedural changes to current operations. These new procedures are developed using one or more wake sensors; however, no additional operational sensors are deployed. Near-term implementations include: Simultaneous Offset Instrument Approach (SOIA) procedure now used at San Francisco International Airport (SFO); a simultaneous staggered approach procedure at Lambert-St. Louis International Airport (STL) that recently received operational approval; and planned extension of the procedure at STL to other airports with similar runway geometries.

Mid-term efforts seek to enhance airport capacity through development/implementation of new weather-sensitive procedural alternatives. Procedures/tools will be developed that factor in crosswind predictions, possibly involving deployment of new meteorological sensors. However, deployment of operational wake sensors is not planned. Specific mid-term efforts underway are addressing reduction of the 2- or 3-min wait for departures behind a Heavy category aircraft from a Closely-Spaced Parallel Runway[*] (CSPR) and reduction of the 2-/3-min wait behind a Heavy from the same runway.

Far-term efforts will investigate solutions involving active monitoring and prediction of wake behaviors, to mitigate the adverse effect of wake turbulence on airport operations. This approach is consistent with the Joint Planning Development Office (JPDO) vision of dynamic separation based upon prevailing atmospheric and traffic conditions. It is likely that this would involve the deployment of one or more wake sensors near each runway end where a capacity increase is sought. It is important to understand that a real-time wake sensor would (a) almost surely be

---

[*] CSPR is defined herein as parallel runways having less than 2,500 ft centerline separation, because, in this regime, wake-based aircraft separation rules may (depending upon the operation and meteorological conditions) govern approaches and departures.

required to satisfy considerably more stringent standards than do current research sensors, and (b) have to be supported by a validated prediction system that would forecast wake behavior over an interval consistent with the time needed to change traffic flow. However, if proved safe and effective, far-term solutions involving real-time wake monitoring would very likely provide greater benefit, albeit at significantly higher cost, than near- and mid-term solutions.

1.2 Passive Acoustic Wake Sensing

1.2.1 SOCRATES Project Background

The Sensor for Optically Characterizing Remote Atmospheric Turbulence Emanating Sound (SOCRATES) is an opto-acoustic based wake-sensing technology that uses laser beams as linear microphones to passively detect acoustic emanations from wakes. From its initial funding in 1997, SOCRATES has been, almost exclusively, a Congressionally-mandated system development undertaking by Flight Safety Technology, Incorporated (FST), New London, Connecticut, with major support from Lockheed Martin Corporation (LMCo), Naval Electronics and Surveillance Systems, Syracuse, New York. Total appropriated funding designated for SOCRATES has been $34.5M during FY1997 to FY2005 (Table 1-1).

Table 1-1 SOCRATES Appropriations History
Source: Library of Congress

Fiscal Year	FAA*	NASA*	Total
1997	$1,589,000	—	$1,589,000
1998	$3,000,000	—	$3,000,000
1999	$3,000,000	—	$3,000,000
2000	$3,000,000	$2,000,000	$5,000,000
2001	—	$2,000,000	$2,000,000
2002	—	$5,000,000	$5,000,000
2003	—	$5,000,000	$5,000,000
2004	—	$4,970,000	$4,970,000
2005	—	$4,960,000	$4,960,000
2006	—	—	—
2007	—	—	—
TOTAL	$10,589,000	$23,930,000	$34,519,000

* Includes effects of rescissions

For most of the SOCRATES project, the Department of Transportation (DOT) Research and Innovative Technology Administration (RITA) Volpe National Transportation Systems Center (Volpe Center) provided technical and project management support to the appropriated agencies. Under contract to the Volpe Center, FST, teamed with LMCo, developed and field-tested a series of prototype sensors.

Two fundamental premises underlie the SOCRATES concept:
- Phenomenology Premise: Potential atmospheric hazards, including wake vortices, emit sound; and

- Sensor Premise: An array of SOCRATES laser beams can remotely (i.e., without directly interacting with the wake flow field) detect, and the associated signal processing computers track, the sound these hazards produce.

This report addresses the validity of these premises in Chapters 1 and 4, respectively.* Additional information on the SOCRATES history and earlier field tests, written from the perspective of the FST-LMCo team, can be found in Ref. 6.

1.2.2 State of Passive Acoustic Wake Sensing Prior to 2003

Although observations of wake vortex sound had been informally reported prior to the SOCRATES project (Refs. 7 and 8) and the investigation of passive wake acoustics using phased microphone arrays was discussed as early as the 1970s (Ref. 9), no systematic effort had been made to explore the intrinsic phenomenology and its operational implications (e.g., are the acoustic emissions consistent enough for wake vortex tracking) until the SOCRATES project. Some efforts were made to measure the wake vortex sound with a single directional microphone to better understand sound emitted by tornados (Refs. 8 and 10), but the details were not circulated in the open literature. These measurements suggested the existence of abundant wake acoustic energy within the 0-200 Hz frequency band.

Three-element microphone measurements were made during the initial field trial of SOCRATES. One of the most salient features reported was a broadband hump in the one-third octave spectra centered at approximately 50-60 Hz for a B-747 (Ref. 11). Informal field observations most often reported wake vortex sound to be either crackling or whistling in nature. A separate study (Ref. 12) argued that the most noticeable features of wake vortex sound was energy in the infrasonic regime (less than 20 Hz), while the SOCRATES array has been designed for higher frequencies, 270 to 360 Hz.

SOCRATES testing prior to 2001 at John F. Kennedy International Airport (JFK) and Langley Air Force Base (AFB) resulted in collection of the then largest known amount of passive vortex acoustic data. However, the general wake vortex science community in 2001 felt that data from the SOCRATES sensor were difficult to interpret because there was insufficient understanding of both the basic scientific phenomenon of wake acoustic generation as well as the SOCRATES sensor's self-noise and its response to a sound source. There were disagreements in the research community regarding the spectral characteristics of wake vortex sound. Furthermore, the premise that aircraft vortices emit a unique and consistent acoustic signature that could be utilized as the basis of a wake vortex sensor was not widely accepted by the research community.

Because of these concerns, a report (Ref. 1) was issued by the Volpe Center in 2001 with the following recommendations:

Recommendation 1: Any further development of SOCRATES sensor technology and other candidate systems dependent upon the hypothesized wake vortex sound-generation phenomenon should be deferred until such time as there is a strong, well-established phenomenological basis for their further development.

Recommendation 2: If there is any interest by the government in pursuing phenomenological research, that is, determining whether wake vortices emit unique, consistent

---

* Terminology used for organization of this report: Chapter (highest level of "indentation"), Section (second highest level of "indentation"), and Subsection (third highest level of "indentation").

acoustic signatures, for possible application to wake vortex detection, the investigation should be initiated and conducted prior to the development of any potential passive acoustic wake vortex sensor.

Recommendation 3: The FAA and NASA should cooperatively develop a set of requirements for an operational wake vortex sensor to support the development and deployment of a wake turbulence system for mitigation of wake turbulence constraints on airport capacity.

1.2.3 Denver 2003 and 2005 Wake Acoustics Tests

Around the same time that these recommendations were made, an independent development in wake acoustics research took place in Europe. The *Deutsches Zentrum für Luft- und Raumfahrt* (DLR, the German Aerospace Center) Institute of Propulsion Technology in Berlin (DLR-Berlin) examined aircraft wake vortex sound under the European Union's C-Wake project. A 64-element microphone array was used to study the feasibility of detecting and tracking aircraft vortices via a passive wake acoustics approach. Two tracks of noise spots were discernable after aircraft flyby, and their motion and spacing were consistent with wake vortex physics (Ref. 13). The DLR study utilized only a very small sample of flights, and the array gain was low with respect to revealing wake noise. Nevertheless, the reported experience and the presentation of results provided the rationale and confidence for using the phased microphone array technique, a better understood technology than use of multiple lasers as lines of equivalent microphones, to conduct a more comprehensive study on aircraft wake vortex sound.

To address the above-listed concerns, and utilizing knowledge gained from the DLR test, the Volpe Center conducted a fundamental wake acoustics test in August and September of 2003 under NASA sponsorship (Ref. 14). Denver International Airport (DEN, Figure 1-3) was selected as the test site, for several reasons: high traffic level, a significant fraction of the fleet large enough to have strong wakes, located away from the city in quiet rural environment, ample space available for equipment, and cooperative airport authority and local FAA personnel. The equipment were deployed north of the airport, and were positioned to capture data from aircraft approaching Runway 16L (whose centerline runs almost exactly north-south geographically, although not magnetically).

Wake sensing by passive acoustics is simplest to implement when the sensors are placed directly beneath the aircraft course of flight. The propagation paths are then primarily vertical, and are only minimally changed by vertical variations in the propagation speed. Both sensor performance and safety requirements necessitate that sensors placed beneath the flight path be (depending upon their height) at least several hundred feet, and likely several thousand feet, from the runway threshold. At such locations, wakes will be generated in the OGE regime.

The DEN03 test objective was to utilize a large 252-microphone phased array for establishing the scientific foundation of wake acoustics. Additional sensors included two wake vortex LIght Detection And Ranging (LIDAR) devices serving as the ground truth and a suite of meteorological sensors for characterizing atmospheric conditions. The DEN03 test also provided an opportunity to examine improvements made to the SOCRATES sensor relative to the prototypes deployed at JFK and Langley AFB. A four-beam SOCRATES prototype system was fielded, along with a smaller phased microphone array from DLR under contract to FST.

The DEN03 wake acoustic test data have been analyzed by a number of organizations, including NASA LaRC, Volpe Center, OptiNav, Florida Atlantic University, FST and LMCo. More

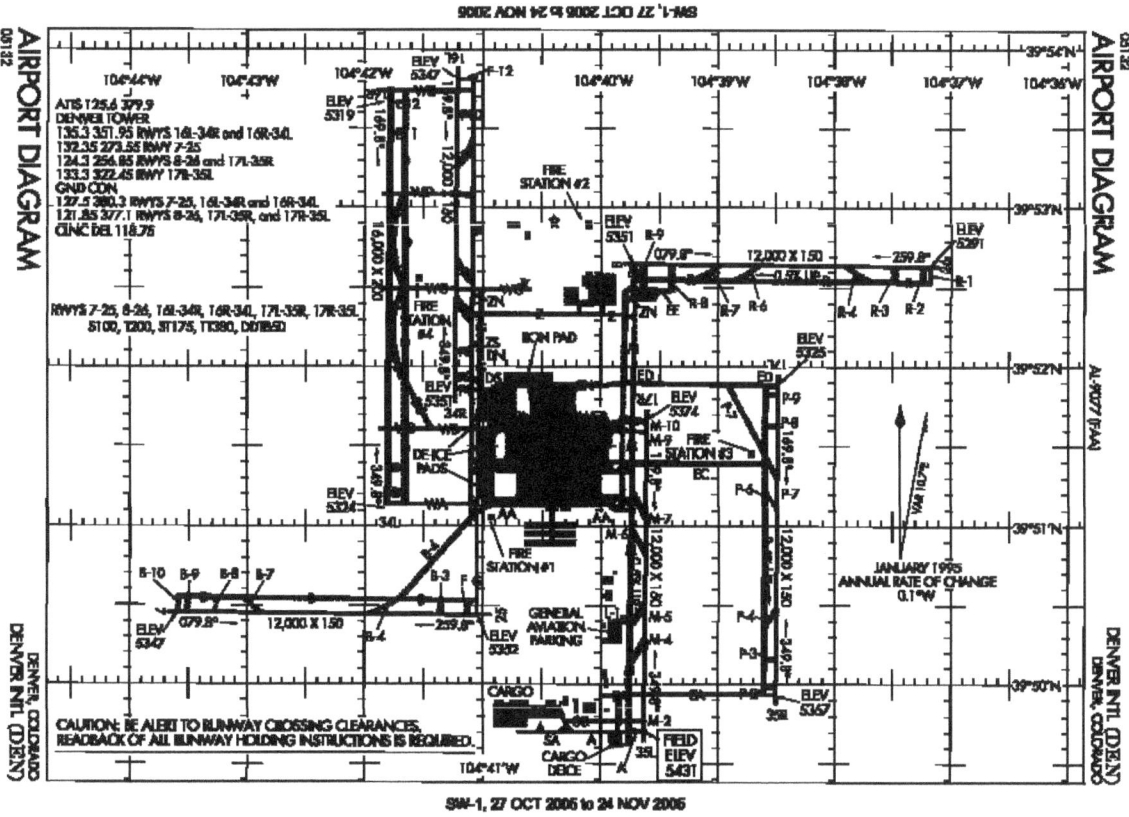

Figure 1-3 Denver International Airport Diagram

recently, Kansas State University embarked on a computational aeroacoustic (CAA) simulation effort based on the data. DLR-Berlin has also continued to analyze their data and performed associated CAA activities. The technical output of the DEN03 test includes a dedicated workshop on wake acoustics as well as numerous papers and reports (e.g., Refs. 15-23).

In the summer of 2005, a 16-beam SOCRATES array, arranged in two configurations, was tested at DEN, again to capture data from aircraft approaching Runway 16L. A Pulsed LIDAR served as the ground truth sensor, and a suite of meteorological sensors was also deployed. The Volpe Center's main focus for the 2005 test was characterization of a specific type of atmospheric fluctuation, Refractive Index Variability (RIV), which may significantly impact SOCRATES performance. A second interest was investigating the impact of atmospheric propagation, which could be an issue if the SOCRATES sensor were to be operated in a mode that listens for a noise source from a standoff-location rather than directly beneath the flight path.

The combination of the DEN03 and DEN05 tests and subsequent analysis efforts have substantially advanced the basic knowledge of wake acoustics as well as of the impact of the RIV noise on SOCRATES performance. It is possible that the better understanding of the RIV noise gained from this work could lead to schemes for alleviating its effects — e.g. improved baffling of the array or the imposition of operational constraints.

1.3 Report Overview

This report is intended to provide a summary of wake acoustic detection capabilities/issues, including development of the SOCRATES sensor, since 2003. Chapter 1 (this one) is introductory in nature, providing background on aircraft wakes and their acoustic emissions as well as an overview of the SOCRATES project. Chapter 2 describes the DEN03 and DEN05 tests which were conducted to advance the state of knowledge of wake acoustics. Chapter 3 presents analysis results which address the fundamental premise that aircraft wakes generate sound and the nature of that sound. Chapter 4 presents an assessment of the capabilities of the SOCRATES prototype sensor deployed at DEN05. Chapter 5 contains a summary of findings and conclusions drawn from the findings.

2. Descriptions of DEN03 and DEN05 Tests

The goals and equipment configurations for the DEN03 and DEN05 tests are described in Section 2.1 and Section 2.2, respectively.

2.1 DEN03 Test Description

Purpose — As stated in Chapter 1, the DEN03 test had as its primary purpose an independent scientific investigation of wake acoustic phenomenology using a 252-microphone array. This goal was supported by other wake and meteorological sensors. Measurements were restricted to aircraft arrivals, because: (a) competing aircraft noise sources (e.g., from engines) are relatively low during this flight phase; and (b) arriving aircraft follow predictable flight paths, enabling instruments to be focused on a fixed location. The DEN03 test also provided the opportunity for the SOCRATES contractor team to conduct developmental testing of their then newest prototype SOCRATES system in the presence of other (both commercially-available and prototype) research instruments which could be used for crosschecking and validation.

Synopsis — Figure 2-1 is an aerial photograph of the test site. The sensors/equipment locations are detailed in Table 2-1. The on-site availabilities of the various DEN03 sensors are shown in Figure 2-2. The sensors are described individually following this Synopsis. The weather sensors, as well as the SODAR and Pulsed LIDAR, did not require operators to collect data. The CW LIDAR and all of the passive acoustic sensors were attended throughout the test.

Figure 2-1 Aerial Photograph of the Denver 2003 Test Site.
Numbered sites are described in Table 2-1.

NASA-DOT Phased Microphone Array — The 252 microphone array, together with its associated recording system, served as the principal instrument of the 2003 test (Figure 2-3) and operated during the period September 3-19, 2003. The center of the array was located 10,812 ft north of the Runway 16L threshold on the runway extended centerline (C/L). The glide slope altitude for this location is 619 ft relative to the threshold, or 709 ft relative to the local terrain. The array system's main components were: 252 electret condenser microphones (Panasonic model WM-61A), eight 32-channel preamplifiers with software configurable gain (made by ACB Engineering), 32 eight-channel 16-bit analog-to-digital (A/D) cards with on-board analog

Table 2-1 DEN03 Sensors/Equipment and Performing Organizations

Location	Sensor / Equipment	Performing Organization(s)
1	Microphone Array	NASA LaRC-DOT Volpe Center
2	Meteorological Tower	DOT Volpe Center
3	SODAR	DOT Volpe Center
4	Instrumentation Trailer, Temperature Profiler and Sonic Anemometer	NASA LaRC - DOT Volpe Center
5	Continuous Wave LIDAR	MIT Lincoln Laboratory
6	SOCRATES Laser Array	FST – LMCo
7	Microphone Array	DLR – Berlin
8	Instrumentation Trailers	FST - LMCo - DLR
9*	Pulsed LIDAR	CTI

*Location 9 is not shown in Figure 2-1; it was 2,931 ft west of Location 1

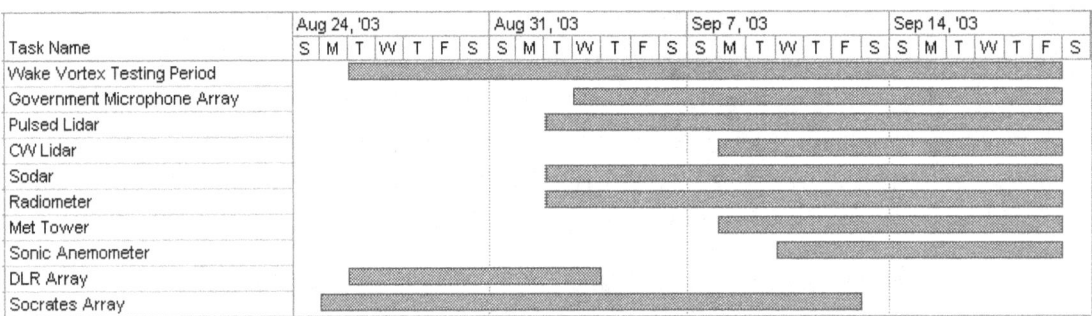

Figure 2-2 DEN03 Test Sensor Availability Timelines

fourth-order Bessel Anti-Aliasing Filters (AAFs) as well as on-board Digital Signal Processors (DSPs, made by Microstar Laboratories), and a data collection workstation (WS).

The microphone nominal sensitivity was -35 dB relative to 1 V/Pa; its rated frequency range was 20-20,000 Hz. However, a calibration test showed that the response was sensitive down to at least 10 Hz. Both the microphones and the preamplifiers exhibited flat frequency responses over the array design frequency band (20-1,000 Hz). The acoustic signal sensed by each microphone was first amplified, then passed through an AAF before being sampled at 153,600 Hz and digitized by the A/D card. The digital signals underwent two stages of filtering using symmetric linear-phase Finite Impulse Response (FIR) filters implemented by the on-board DSP before being down-sampled to 25,600 samples/sec. All the A/D channels were synchronously sampled, which is critical for the implementation of array processing algorithms during post-test analysis.

The microphone placement pattern, designed by OptiNav, is shown in Figure 2-4. The array dimensions were approximately 102 ft longitudinally (parallel to the flight path) by 397.5 ft laterally (perpendicular to the flight path). The array was centered on the extended centerline of Runway 16L. The pattern was designed to provide 3-dB resolutions of 30 ft laterally and 100 ft longitudinally at the glide slope altitude for frequencies greater than 50 Hz while also maintaining a high main-lobe to first side-lobe ratio (15 to 20 dB). The array lateral dimension was significantly larger than its longitudinal dimension to ensure the ability to resolve the two vortices.

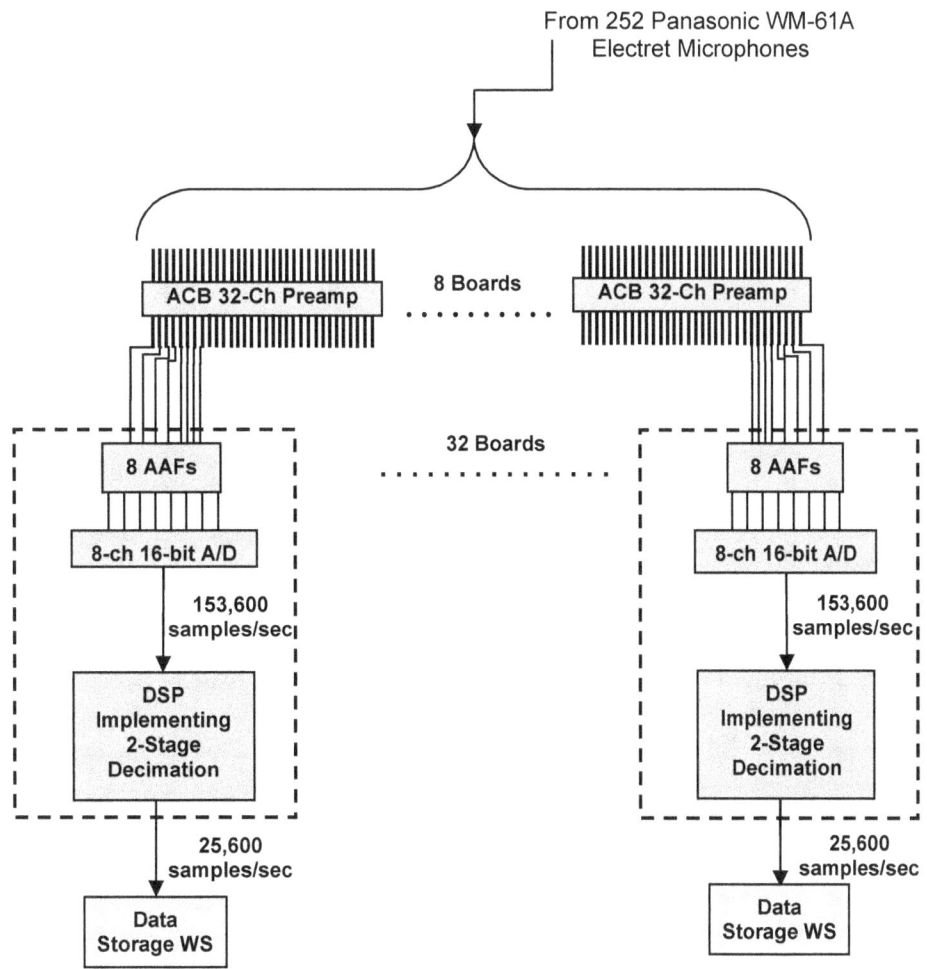

Figure 2-3 NASA-DOT Microphone Array System Diagram

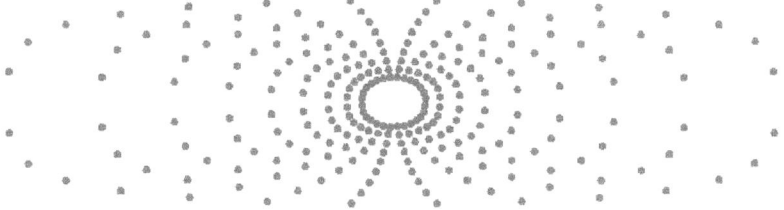

Figure 2-4 NASA-DOT Microphone Array Geometry

Non-uniform (rather than uniform) microphone spacing was chosen to prevent the occurrence of grating lobes[*] while providing high spatial resolution over the 50-1,000 Hz frequency band of interest. The 252 elements forming the array provided an ideal gain of 24 dB when all microphones were weighed equally. The array Point Spread Function (PSF), or response pattern to a monopole source, is shown for the lateral (Figure 2-5) and longitudinal (Figure 2-6) direction for

---

[*] In the antenna response pattern, grating lobes are undesired identical copies of the main beam which arise at frequencies greater than the upper design frequency (here, 1,000 Hz).

two frequencies (100 and 340 Hz) and two aircraft heights (200 and 500 ft AGL). Table 2-2 tabulates the array 3-dB resolution for the same heights and frequencies.

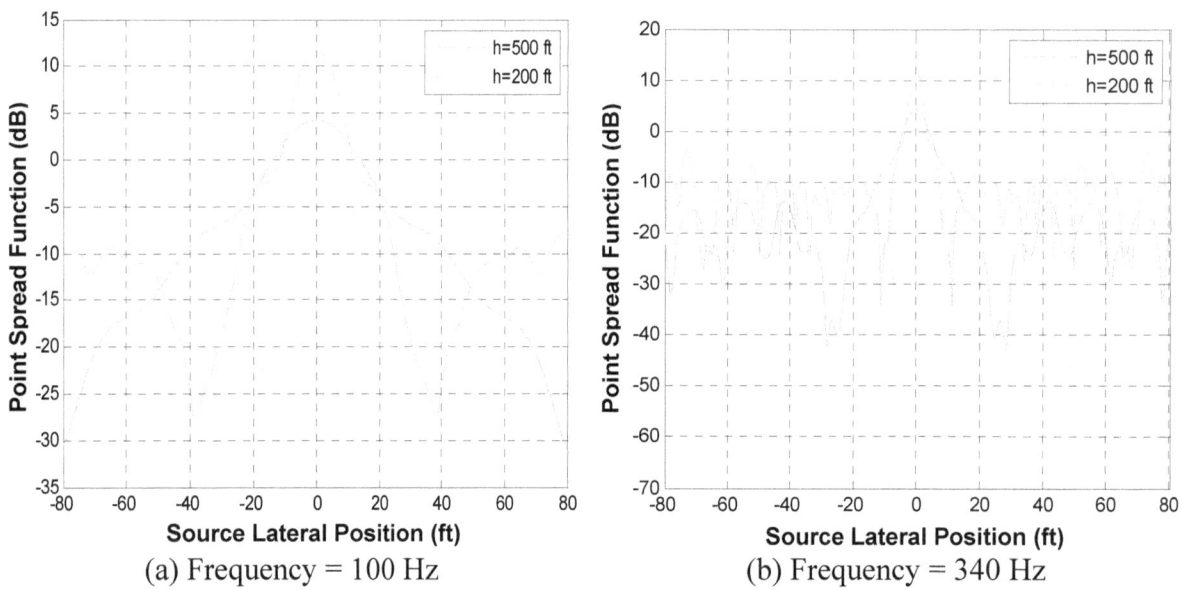

(a) Frequency = 100 Hz   (b) Frequency = 340 Hz

Figure 2-5 Microphone Array PSF in Lateral Direction for Two Frequencies

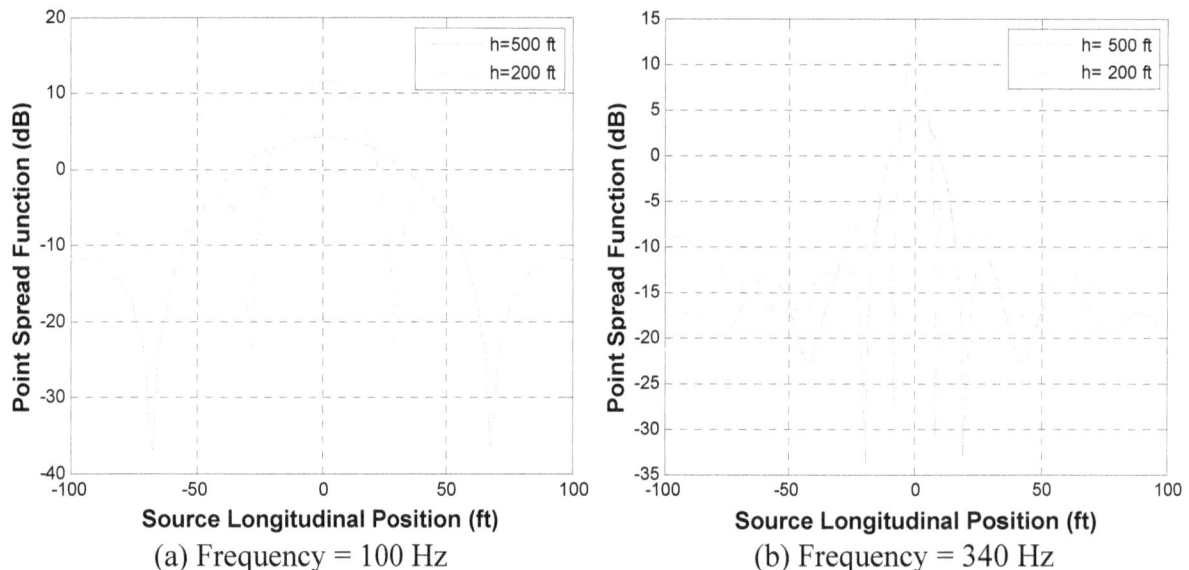

(a) Frequency = 100 Hz   (b) Frequency = 340 Hz

Figure 2-6 Microphone Array PSF in Longitudinal Direction for Two Frequencies

Table 2-2 Microphone Array Resolution

Source Height \ Source Frequency	Lateral		Longitudinal	
	100 Hz	340 Hz	100 Hz	340 Hz
200 ft	11 ft	3.5 ft	25 ft	8 ft
500 ft	22.5 ft	6.5 ft	57 ft	17 ft

For each aircraft arrival, acoustic data collection started a few seconds before the aircraft flew by the array; the data collection duration was up to 90 sec. Results of that processing are the basis of Chapter 1. Further information on the microphone array can be found in Refs. 15-16.

The microphone array measurement dataset initially consisted of 1,011 flybys. After excluding files with late or false triggering, aircraft departures, and files that were too short to be meaningful, the array dataset used for analysis had a total of 771 arrivals. Approximately 99% of the measurements were made under Visual Meteorological Conditions (VMC). Aircraft types ranged in size from Beech 1900 to Boeing 777 (Figure 2-7).

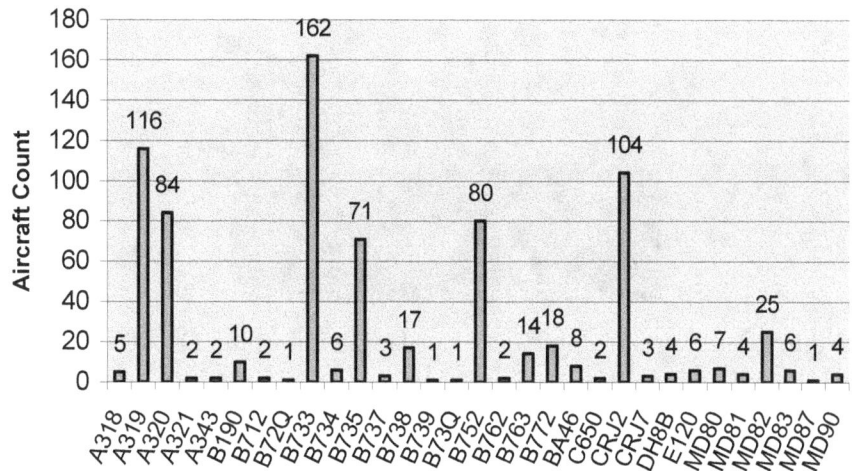

Figure 2-7 Aircraft Type Distribution for the Microphone Array Data Set

Meteorological Tower Sensors[*] — Weather conditions strongly influence wake vortex behavior. A suite of weather sensors was therefore fielded to support both wake aerodynamics studies and acoustic data processing. The meteorological sensors included an instrumented 107-ft tower (Figure 2-8) as well as others to be discussed later. The tower was intended to measure the portion of the wind profile closest to the ground, and was instrumented with three-axis R. M. Young Gill propeller anemometers at three heights and a Vaisala temperature and relative humidity sensor near the mid-height of the tower.

SODAR — An AeroVironment SODAR was used to measure the wind field above the height of the meteorological tower. The SODAR had three 4,500 Hz acoustic beams that transmitted and received cyclically in each 1-sec interval. It was deployed to provide three-component wind measurements from 35 m up to 200 m (approximately 115 ft to 660 ft) at 5 m (approximately 16 ft) resolution. The maximum measurement height actually achieved varied, depending upon the prevailing atmospheric conditions.

NASA-DOT Instrumentation Trailer, Temperature Profiler, and Sonic Anemometer — The trailer housed the control and backup computers for the NASA-DOT microphone array. Also mounted on the trailer were a Metek ultrasonic anemometer and a Kipp & Zonen MTP5 microwave radiometer passive temperature profiler. The ultrasonic anemometer was used to

---

[*] Consistent with their scientific origins, the DEN03/DEN05 instruments often utilized the metric system for measurement and recording. However, the descriptions and sample results presented herein generally include the equivalent English units, as these are standard for aviation.

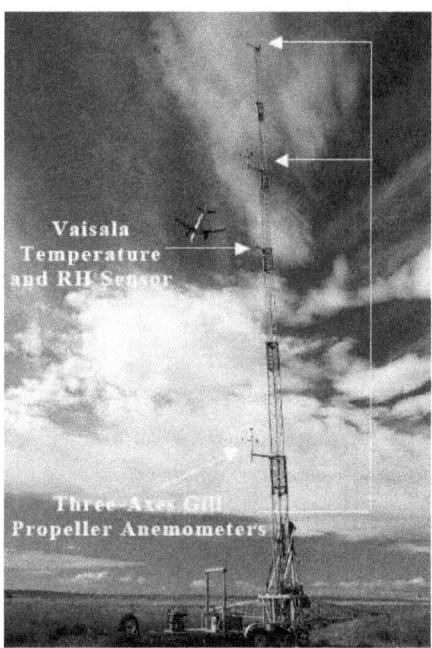

Figure 2-8 Meteorological Tower Deployed in the DEN03 Test

provide a point measurement of the atmospheric turbulence characteristics — e.g., Turbulent Kinetic Energy (TKE) and Eddy Dissipation Rate (EDR) — as well as temperature.

The microwave radiometer measured the thermal radiation of oxygen molecules to obtain temperature profiles from the ground to 600 m (approximately 1,980 ft) over the array area. The temperature profile was collected to characterize the stratification of the atmosphere as well as acoustic propagation. Additional weather information — such as barometric pressure, ceiling, and Runway Visual Range (RVR) — was obtained from the airport Automated Surface Observing System (ASOS).

LIDAR Systems — Two Doppler LIDARs were deployed as ground truth wake track sensors during the DEN03 test: a Coherent Technologies, Inc. (CTI, now part of Lockheed Martin Corp.) Pulsed LIDAR and a Massachusetts Institute of Technology Lincoln Laboratory (MIT-LL) CW LIDAR. Both were located west of the test site, and are not visible in Figure 2-1. Both LIDARs measured the component of the flow field velocity vector along the line-of-sight between the sensor and the measurement volume. This was accomplished by detecting the Doppler shift of the light backscattered from the aerosols naturally present in the atmosphere.

The CW LIDAR emitted a continuous beam of 10.6 μm (approximately 0.000,42 in.) wavelength and collected data by adjusting the focus range of the laser. By varying the elevation angle, the beam was scanned across the region of interest. Because of the continuous nature of the LIDAR, detailed velocity profiles through the wake could be obtained.

The CTI Pulsed LIDAR (Figure 2-9) used a 2 μm (approximately 0.000,079 in.) wavelength laser which was pulsed at a rate of 500 Hz. Like pulsed radio frequency radars, it measured distance by range-gating the returned signal. Both the CW and Pulsed LIDARs provided wake position and strength data. The Pulsed LIDAR was also used to obtain crosswind profiles from wake vortex scans, as well as dedicated three-component wind velocity from 200 ft to (when weather permitted) 2,200 ft AGL from conical scans.

Figure 2-9 CTI Pulsed LIDAR Mounted on Trailer

SOCRATES Sensor — The 2003 test provided an opportunity to field test the latest SOCRATES prototype sensor. A four-beam SOCRATES array (Figure 2-10) was deployed under the extended centerline of Runway 16L, and operated during the period August 25 to September 12, 2003. Each of the SOCRATES laser beams was 50 m (164 ft) long and was elevated approximately 3.5 ft above local ground. The beams were oriented parallel to the extended runway centerline, and were spaced 18 in. apart.

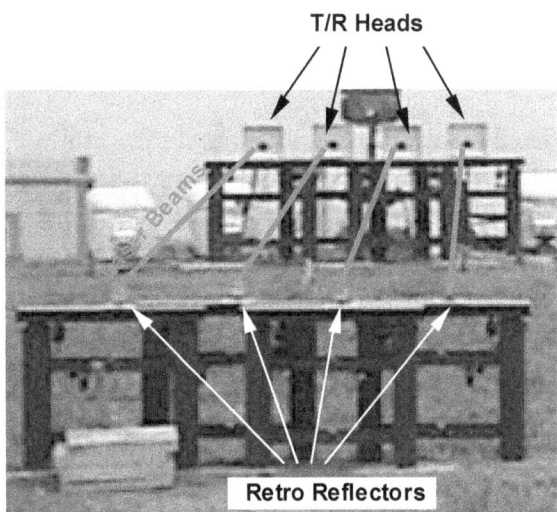

Figure 2-10 Four-Beam SOCRATES 2003 Array in the Skyward-Looking Configuration

2.2 DEN05 Test Description

Purpose — The DEN05 test had two main purposes, and was separated into two corresponding measurement phases: (1) developmental testing by FST and LMCo of a new 16-beam prototype SOCRATES sensor (September 9 to October 12, 2005); and (2) atmospheric measurements by the Government to characterize RIV noise, a significant factor in SOCRATES performance (December 15 and 16, 2005).

Synopsis — As was the case in the DEN03 test, measurements were collected under VMC conditions more than 99% of the time. The airplane distribution during times when the SOCRATES Skyward-Listening array was operational is shown in Figure 2-11.

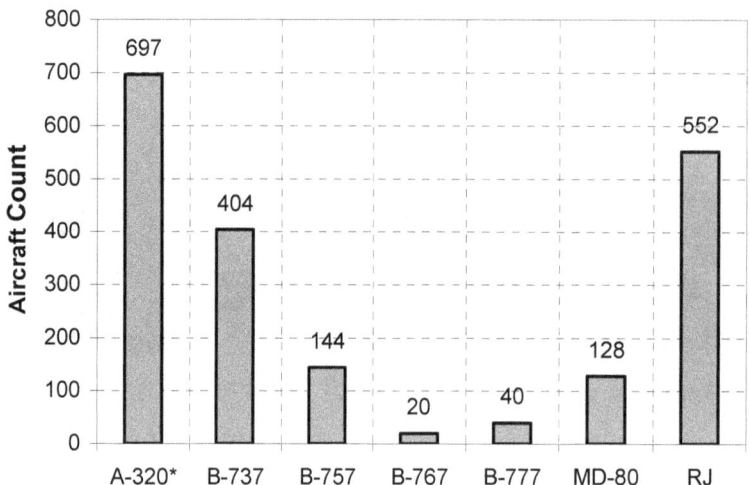

Figure 2-11 Aircraft Observed by DEN05 SOCRATES Skyward-Listening Array
*In the FST-LMCo analyses, the "A-320" category consisted of four Airbus aircraft models: A-318, A-319, A-320 and A-321.

Figure 2-12 depicts the site layout for the DEN05 test, which utilized approximately the same area north of DEN airport as the 2003 test. Runway 16L arrival traffic was again utilized for the test. The sensors/equipment locations are detailed in Table 2-3. Availabilities of the sensors are shown in Figure 2-13.

Table 2-3 DEN05 Sensors/Equipment and Performing Organizations

Location	Sensor / Equipment	Technical Organization(s)
1	Meteorological Tower	DOT Volpe Center
2	Instrumentation Trailer, Temperature Profiler	NASA LaRC - DOT Volpe Center
3	SOCRATES Laser Array in the Skyward-Looking Configuration	FST - LMCo
4	SOCRATES Laser Array in the Billboard Configuration	FST - LMCo
5	Hotwire and Cold Wire Sensors	DOT Volpe Center
6	Pulsed LIDAR (Site 1)	CTI
7	Pulsed LIDAR (Site 2)	CTI

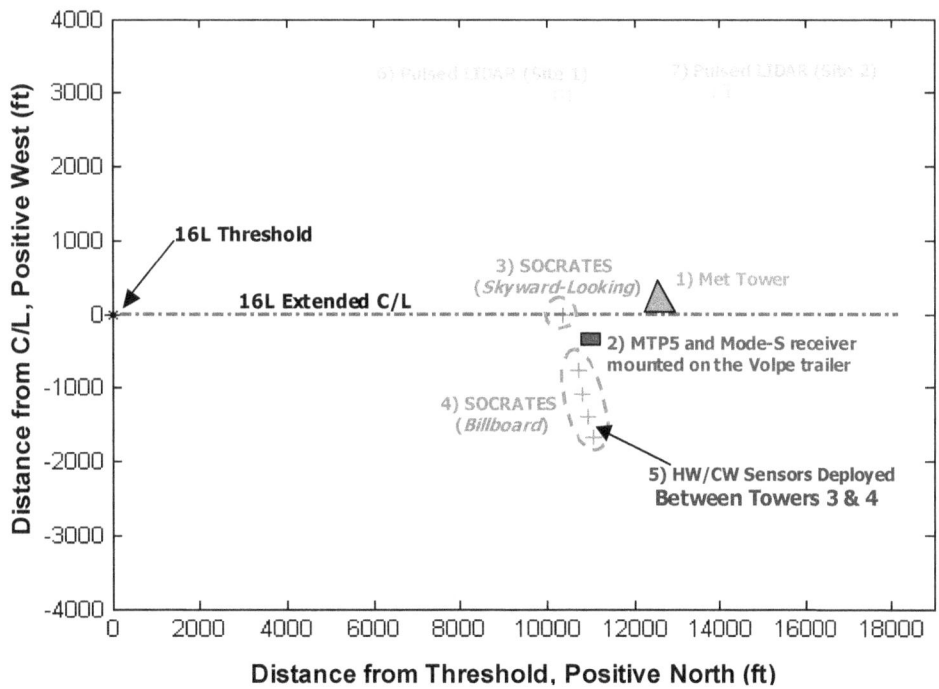

Figure 2-12 Plan View Diagram of the DEN05 Test Site
Numbered sites are described in Table 2-3

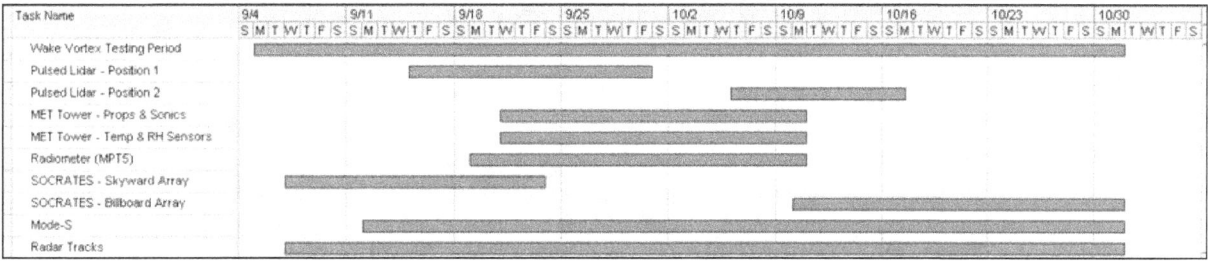

Figure 2-13 DEN05 Test Sensor Availability Timelines

SOCRATES — The 2005 field trial was significant in that the FST-LMCo team considered the campaign to be a full system test. This was the first SOCRATES test employing 16 laser beams, and involved two separate beam configurations.

- For the Skyward-Looking configuration, the beams were deployed in two eight-beam sub-arrays, arranged horizontally, one on either side of (and oriented parallel to) the runway centerline. This arrangement enabled vertical and lateral (i.e., orthogonal to the extended centerline) tracking of wake locations. A sample arrangement is shown in Figure 2-14(a).

- For the Billboard configuration the laser beams were vertically stacked on concrete towers for the purpose of standoff listening for wakes near the start of the stabilized approach path approximately 1,000 m (approximately 3,280 ft) away from the sensor. A photograph of the Billboard array is shown in Figure 2-14(b).

Further discussions as well as simulation results for the SOCRATES array are presented in the SOCRATES system analysis and evaluation section of Chapter 4.

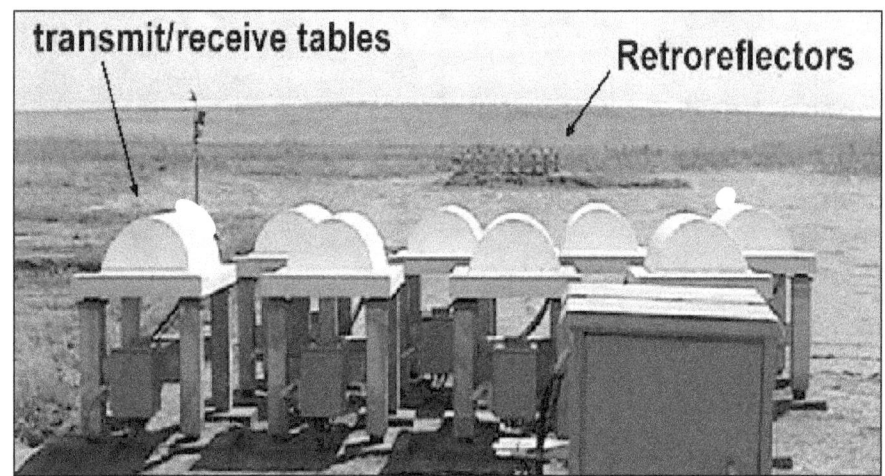
(a) One of Two Skyward-Looking Sub-arrays

(b) Billboard Array during Construction

Figure 2-14 Photographs of Two DEN05 SOCRATES Configurations

Pulsed LIDAR — A Pulsed LIDAR was deployed as the "ground truth" sensor for providing wake position and strength information. As was in the case of 2003, the Pulsed LIDAR (Figure 2-9) was scanned in a plane perpendicular to the extended runway centerline so that crosswind profiles were also obtained. In addition, the Pulsed LIDAR conducted conical Velocity Azimuth Display (VAD) scans to obtain both headwind and crosswind from 200 ft to 2,200 ft AGL. The wind measurements also were made to support the study of sound propagation variations due to wind and temperature, which become important when passive acoustic sensors listen to a source some distance away. For the Skyward-Looking array, the Pulsed LIDAR was scanned over the SOCRATES focusing volume directly above the sensors. For the Billboard array, the LIDAR was repositioned 1,000 m (approximately 3,280 ft) farther up the glide slope to scan over the focusing point of that array configuration.

Meteorological Tower — The meteorological tower (Figure 2-15) was instrumented with three Campbell Scientific CSAT3 three-dimensional sonic anemometers, three Campbell Scientific FW05 fine wire thermocouples, three Vaisala temperature and relative humidity probes, and nine R.M. Young propeller anemometers with orientations optimized for headwinds. The meteorological tower provided lower-altitude winds to supplement the VAD wind from the Pulsed LIDAR. The sonic anemometer data yielded EDR information up to 106 ft. Through temperature measurements from the tower and the temperature profiler (described next), EDR at higher altitudes could be inferred.

Figure 2-15 Photograph of DEN05 Meteorological Tower

Microwave Radiometer — The same temperature profiler fielded in DEN03 was also deployed in DEN05 to support the evaluation of the influence of atmospheric conditions on the Billboard array's design constraints (Figure 2-16). Temperature data were also used to determine atmospheric stability classes for prediction of wake dynamics.

Hotwire Sensors, Cold Wire Sensors, Sonic Anemometers, and Fine Wire Thermocouples — Hotwire (HW) and Cold Wire (CW) sensors which measure high frequency velocity and temperature fluctuations, respectively, were deployed at the billboard SOCRATES array site, to obtain data for the modeling of RIV noise (discussed as part of the SOCRATES system analysis in Chapter 4). These sensors were deployed at three locations that formed a straight line along the SOCRATES beams (Figure 2-17). As shown in Figure 2-18, the HW/CW sensors were mounted near the measurement volume of a CSAT3 sonic anemometer and close to the tip of a fine wire thermocouple (TC) at each location, to allow for in-situ calibrations of these sensors.

Figure 2-16 Photograph of Microwave Radiometer

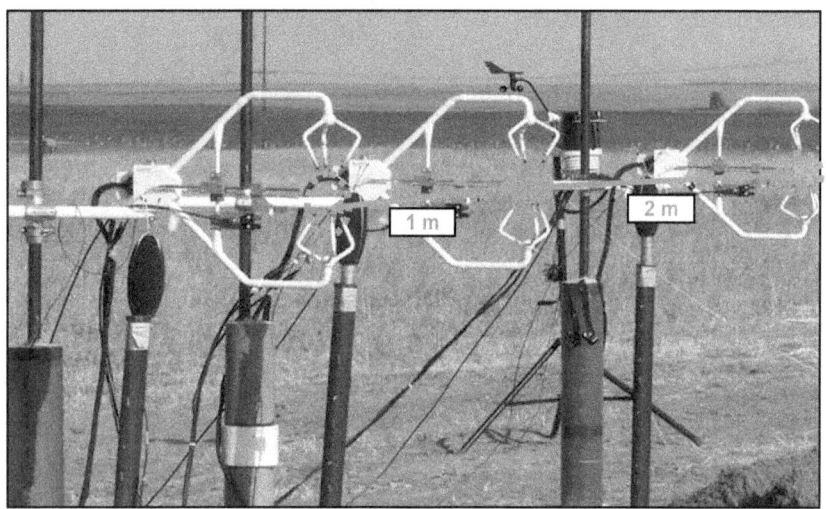

Figure 2-17 Setup for the RIV Noise Experiment

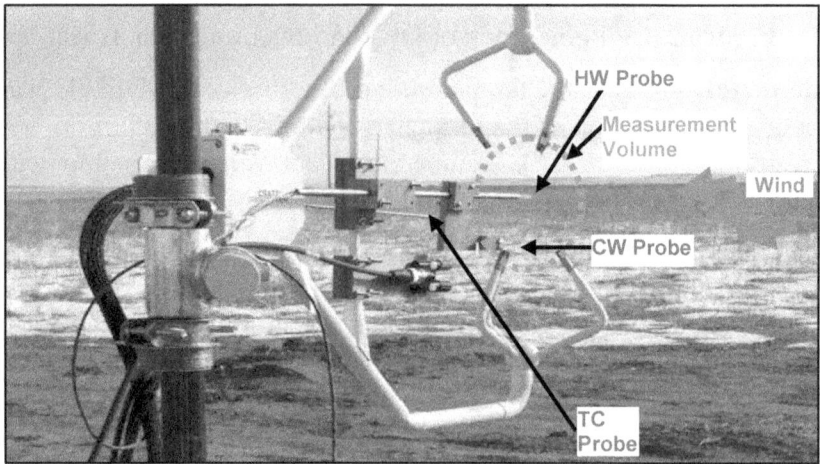

Figure 2-18 Close-Up of the Hotwire, Cold Wire and Thermocouple Probes

3. Results of Denver Tests: Wake Acoustic Phenomenology

This chapter addresses phenomenology issues fundamental to wake detection by passive acoustic means, whether by microphones, SOCRATES or another sensor. Stated as questions, the most important issues are:
- Do aircraft wakes generate sound consistently?
- How well can wake sound be tracked in location and duration?
- How well can wake strength be determined from sound received on the ground?

Section 3.1 first examines in detail the sound generated by specific approaching aircraft as received by the NASA-DOT microphone array during the DEN03 test. Section 3.2 then employs statistics to examine the consistency of wake sound collected from a large set of approaches during DEN03. Section 3.3 compares sound detection by the SOCRATES laser beams with other sensors deployed during the DEN03 test. Section 3.4 examines the frequency distribution of vortex-generated sound. Lastly, Section 3.5 examines the relationship between vortex sound and vortex strength.

3.1 Evidence of Vortex-Generated Sound (NASA-DOT Microphone Array)

This section examines the most fundamental issue — the existence of wake vortex generated sound — using familiar microphone technology. It was first important to establish that, when an aircraft flew by the microphone array, the sound received was in fact from wake vortices rather than from other sources — e.g., engine exhausts, fuselage vibrations, or the local environment. Ideally, the processed microphone data would reveal the existence of two line noise sources behind the aircraft whose spatial and temporal characteristics were consistent with well-established wake vortex behavior. Also, wake vortex tracks derived from acoustic emission should allow comparison with Pulsed LIDAR tracks, in order to establish their consistency. This would be strong evidence of the sound being from the wake vortices, as LIDAR tracks are derived from the tangential velocity flow field of the vortices.

To address this issue, microphone array signals were processed using three-dimensional beamforming[*] techniques[†] (Ref. 16). In brief, when the array of microphones anticipates a signal from a particular location in the sky (i.e., a focusing point), a pattern of the time delays is established that theoretically corresponds to the propagation path lengths from the signal source to the various microphones. In processing, the signal received by each microphone is delayed by an amount corresponding to the difference between its propagation path and the longest path. The delayed signals are then summed. If the signals received by the various microphones match the assumed time delay pattern for the focusing point, then the delayed and summed versions of these signals will be constructively reinforced. By this process, energy from surrounding acoustic sources will be summed incoherently, which has the effect of suppressing unwanted

---

[*] The term "beam" has two distinct meanings herein: (1) SOCRATES laser energy path, introduced earlier, which is always oriented horizontally and located near the ground; and (2) receiving response pattern of an array of acoustic sensors (microphones or laser beams), introduced here, which is directed toward the aircraft wake and is generally oriented with a significant vertical component. When a distinction is needed, the latter is termed an acoustic beam.

[†] Three-dimensional acoustic beamforming assumes that the sound radiates from a point source, so that the wave-fronts are spherical. The more common two-dimensional beamforming technique assumes the sound source is "infinitely far" away, so the wavefronts are planar.

noise. The summed signal amplitudes can then be plotted as color contours, providing a visual representation of the location and relative intensity of the acoustic source distribution.

For examining whether sound behind an aircraft was consistent with wake vortex behavior, the beamforming process was first done in a vertical plane perpendicular to the flight path and directly over the array center, to obtain a time history of acoustic source-height versus lateral position. The vertical beamforming results were then used to construct depictions of horizontal planes at various times to observe the spatial variation of these sound sources behind the aircraft.

Figure 3-1 shows examples from the vertical-horizontal beamforming analyses for arrivals by a B-767 on September 3, 2003, at 07:43 pm local time, and by a B-737 on September 16, 2003, at 11:25 am local time. The beamforming frequency range was 0-300 Hz, and a 5 ft by 5 ft grid of focusing points was used. Each frame is the result of a 2-sec processing (integration) interval, and the number that appears in the upper left hand corner indicates the frame number. For brevity, only every other frame is shown. Aircraft engine and airframe noise are seen in the first beamforming results, and are manifested as elliptical blurs. The silhouette of the aircraft is superimposed onto the frames containing the aircraft flyby noise for reference. The direction of flight is from north to south (in the figure, from the upper to the lower portion of the frame). Between 4 sec and 8 sec after aircraft flyby, two lines of noise sources become evident; moreover, their initial separation distance varies with wingspan (i.e., is larger for B-767 than for B-737). The separation distances of the line sources are in good agreement with their theoretical values of $\pi/4$ times, or 79% of, the aircraft wing spans.

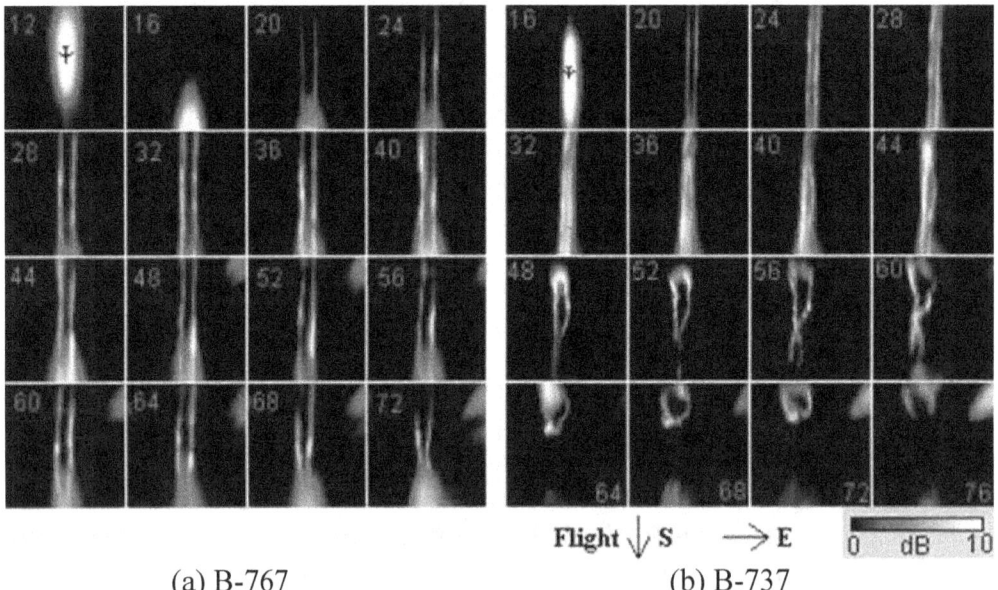

(a) B-767   (b) B-737

Figure 3-1 Examples of Acoustic Noise Source Localization Map Behind Aircraft
Beamforming frequency band is 0-300 Hz; grid of focusing points is 5 ft by 5 ft.

In the case of the B-737 flyby, at frame number of 48, it can be seen that the line vortices have evolved into a ring state, apparently through a process known as Crow instability (whose occurrence can sometimes be revealed in aircraft contrails). Also, the two vortices descended at rates that are in good agreement with their theoretical values (Ref. 15). Thus several aspects of the acoustic data collected by the microphone array were consistent with established wake vortex dynamics.

The ability to resolve both vortices at the expected separation distances for wing-mounted-engine aircraft such as the example shown in Figure 3-1, as well as for a fuselage-mounted-engine — e.g., an MD-80 — whose engine spacing is far shorter than the vortex spacing, is another indication that noise from aircraft vortices have been tracked instead of noise from engine plumes (Ref. 17). Lastly, although size of the oval encompassing both vortices is up to twice the aircraft wingspan, the source localization maps reveal that the vortex sound generation is predominantly from a confined region in or around the vortex cores. Thus, there are several items of evidence indicating that the acoustic sources revealed by the beamforming process are vortices, not engine exhausts or other sound radiating from the aircraft.

A comparison of microphone array wake tracks with Pulsed LIDAR tracks provides further proof that the acoustic sources found behind the aircraft are wake vortices. In Figure 3-2, a B-757-200 (abbreviated B752) flyby on September 3, at 06:39 pm local time, is shown using data from the microphone array vertical beamforming (blue lines) and the Pulsed LIDAR (red lines). The array had a 2-sec sampling interval and the beamforming frequency band is 200-400 Hz; in contrast, the LIDAR acquired track information on a 5-sec interval.

The top half of Figure 3-2 shows the lateral motion of the wake versus time, with zero position being the extended runway centerline and positive values indicating the eastward direction. The lateral position trend shows that a wind from the west was present during data collection. The bottom half of Figure 3-2 shows the vortex height above ground level as a function of time. Recall that microphone data were usually collected up to 90 sec, with the microphones activated as the aircraft approached the array (Ref. 16). Some of the vortex tracks derived from the array are thus artificially shortened, as is thought to be the case here. In general, the lateral position comparison between the two sensors is better than the vertical position comparison; the latter is related to the poorer resolution of this particular microphone array in the vertical direction.

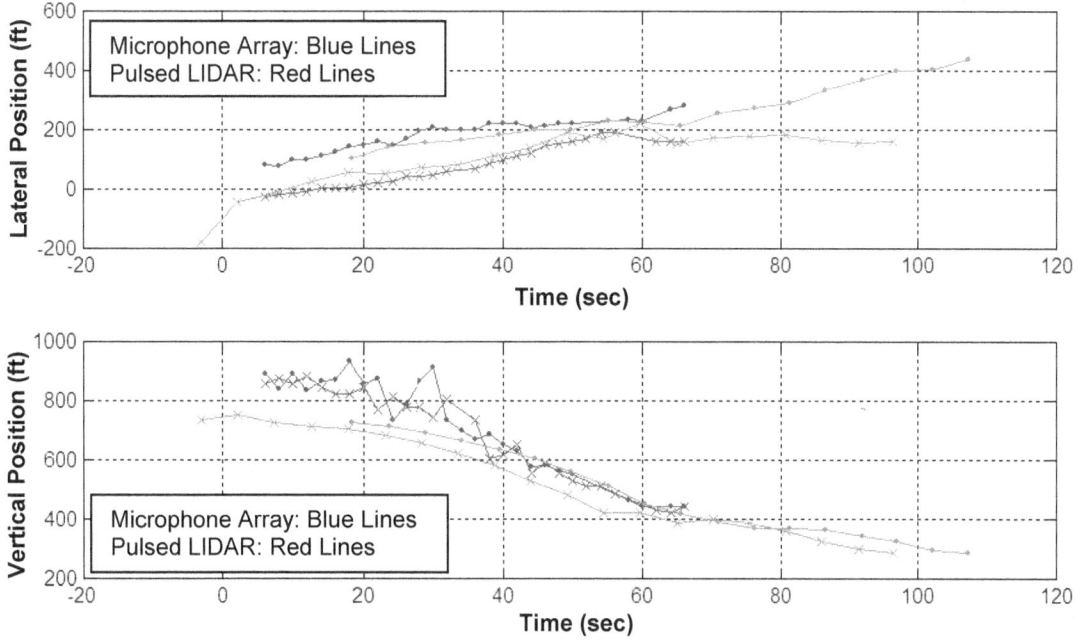

Figure 3-2 Comparison of Vortex Trajectories from Microphone Array and Pulsed LIDAR B-757-200 aircraft; 200-400 Hz microphone beamforming frequency band.

In Figure 3-2, the general trajectories of the wake vortices as determined from the two techniques are similar. At 54 sec after flyby, both sensors reported the vortex spacing to be dramatically reduced, a pinch/linking-like behavior. The corresponding horizontal beamforming run for this case confirmed that a Crow linking process is indeed taking place. The post linking state however, is artificially terminated at 66 sec. Processing details are found in Ref. 18.

Analyses such as those presented in Figure 3-1 and Figure 3-2 demonstrate conclusively that aircraft wake vortices can generate acoustic signatures whose behavior is similar to wake behavior derived from flow velocity measured by sensors such as LIDARs. A more global characterization of the comparison between the microphone array and Pulsed LIDAR derived wake tracks is contained in Ref. 19. Additional comparisons between the microphone array tracks derived from a frequency-domain beamforming scheme and both Pulsed and CW LIDARs are found in Ref. 20.

Although not reported in detail herein, a wide variety of processing techniques — including time and frequency domain delay-and-sum beamforming (Refs. 16 and 20) with de-convolution enhancements (Refs. 21 and 22), and more recently, wavelet-based time domain beamforming (Ref. 23) — have been employed to analyze the DEN03 microphone array data. Results from these other processing techniques did not alter the fundamental features seen in the source localization maps such as Figure 3-1, reinforcing the conclusion that the acoustic imagery was due to aircraft wake vortices and was not the result of processing artifacts.

3.2 Consistency of Wake Vortex Sound Generation

While Section 3.1 employs example flybys to demonstrate that wake vortices did generate sound for selected approaches, it's possible that those cases involved known or unknown favorable conditions. This section therefore addresses the consistency of wake sound generation for the entire DEN03 test and three sensors — the NASA-DOT microphone array (Subsection 3.2.1), the DLR microphone array (Subsection 3.2.2), and the SOCRATES 2003 prototype (Subsection 3.2.3) — as well as the DEN05 test and the SOCRATES 2005 prototype (Subsection 3.2.4).

3.2.1 NASA-DOT Microphone Array

Since the high resolution beamforming approach used to generate Figure 3-1 and Figure 3-2 is not well suited to the batch processing needed to characterize the entire DEN03 test data set, a different strategy was implemented. Source localization maps were generated for approaches from September 3 to September 19 using horizontal beamforming for two frequency bands, 0-200 Hz and 200-400 Hz. For both bands, data were processed for a fixed horizontal focusing plane 500 ft above the ground and a coarse grid resolution of 20 ft by 20 ft (interpolated to 2 ft by 2 ft resolution in post-processing) that spanned 1,500 ft in the lateral (east-west) axis and 1,000 ft in the longitudinal (north-south) axis. An integration time of 2 sec was used for each grid point. For each track, the images were converted to movie frames, yielding one movie per track with a 2-sec increment between frames. The resulting 771 movies were independently inspected for evidence of a wake by two reviewers, and the scores were averaged.

Simulation of the array response in the vertical direction at various frequencies of interest yielded results similar to those shown in Figures 2-5 and 2-6; these demonstrated that a fixed beamforming altitude of 500 ft was more than adequate for an analysis of this type. The poor resolution of the array in the vertical direction permitted a constant altitude beamforming such that vortex sound, when emitted with sufficient loudness, could be revealed from aircraft altitudes down to approximately 250 ft above the ground.

An example of horizontal beamforming results for a B-737-300 flyby is presented in Figure 3-3 (in the figure, t=0 corresponds to the aircraft directly above the array). The array spatial resolution is inherently better at higher frequencies, so that the finer definition seen for the 200-400 Hz band is expected. However, the 200-400 Hz images usually have a fainter and spottier appearance, indicating that less acoustic energy is present in that band. (Power spectra presented in Section 3.4 quantify the small amount of wake acoustic energy relative to the background noise for frequencies greater than 100 Hz.) It is, however, important to note that the 0-200 Hz and 200-400 Hz beamforming revealed the same wake vortex behavior when the array gain was sufficient, as is the case here. For example, the columnar form of the vortices become contorted at 36 sec and clearly formed into vortex rings, as is seen in the results for 52 sec after flyby.

As part of this analysis, spectral characterizations of wake vortices from a few randomly selected approaches were made. These characterizations confirmed that 400 Hz is a reasonable upper bound for the analysis, as there is little acoustic energy above that frequency.

Figure 3-4 summarizes the vortex noise detection statistics as revealed by the source localization maps derived from the NASA-DOT microphone array. The figure also shows the degree of wake noise detection consistency by aircraft type and by frequency band (i.e., 0-200 Hz and 200-400 Hz). Several observations can be made:

- The overall detection probability was in the 80% to 85% range.

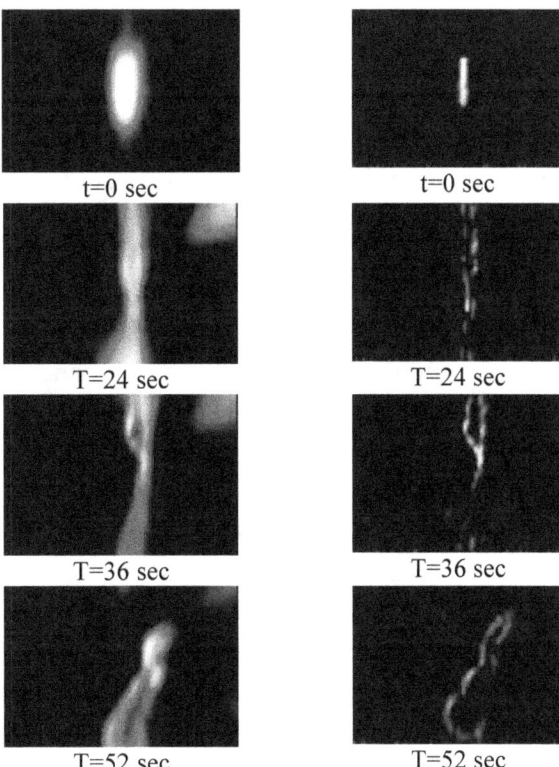

(a) 0-200 Hz Band, 10 dB Dynamic Range   (b) 200-400 Hz Band, 6 dB Dynamic Range

Figure 3-3 Horizontal Beamforming Batch Processing for a B-737-300

The two bands show similar wake dynamics. Energy in the higher-frequency band is noticeably less.

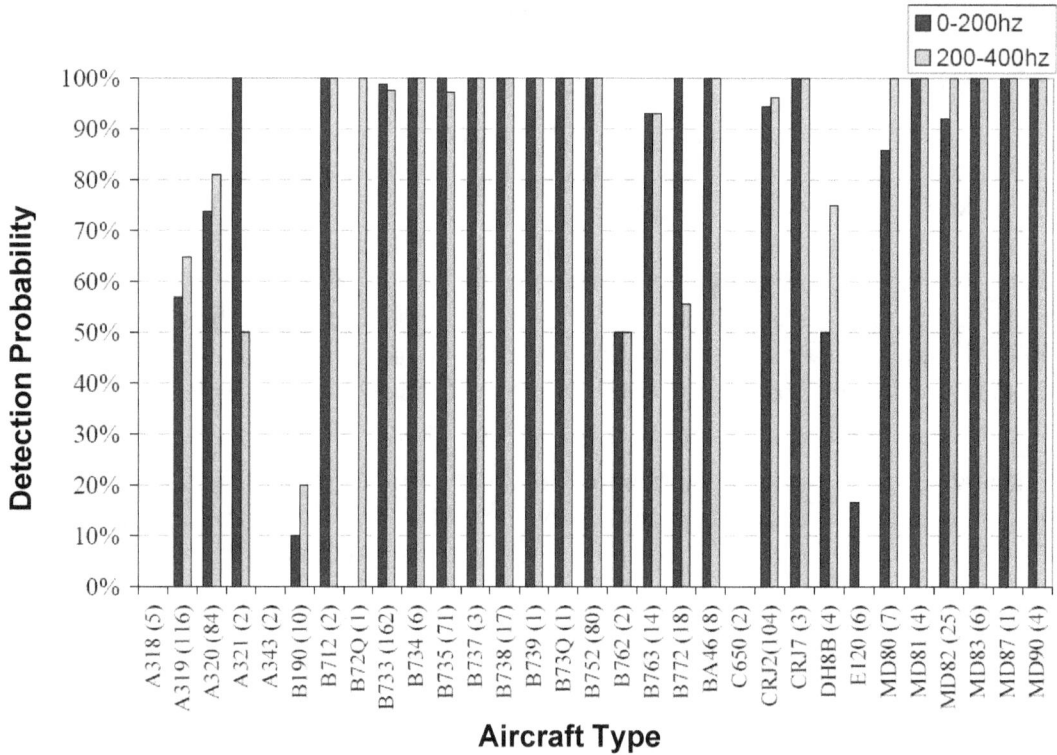

Figure 3-4 Wake Vortex Detection Probabilities for NASA-DOT Microphone Array
The number of approaches is shown in parenthesis after the aircraft model.

- While for some aircraft the two frequency bands had different detection probabilities, neither band had a clearly better overall probability of detection.

- The Boeing 737 aircraft detection rate was over 95% for both frequency bands and all seven series in the database. Similarly, the detection rate was 100% for the 80 approaches by B-757-200 aircraft and both frequency bands.

- Boeing 767 and 777 aircraft detection rates were between 50% and 100% and differed by frequency band. The relatively low B-777-200 detection in the higher band is surprising; however, the associated database is not very large. B-777 aircraft produced consistent wake sound in the lower frequency band.

- Airbus aircraft were generally less well detected than Boeing aircraft. A significant portion of the A-318, A-319 and A-320 aircraft models did not produce detectable wake vortex sound in either frequency band. The second heaviest aircraft in the database is the A-340-300, but the two arrivals in the sample were not detected.

- McDonnell Douglas (MD) manufactured aircraft (six types in database) were generally well detected in both frequency bands.

- For the 104 approaches observed, Bombardier Canadair Regional Jet 200 series (CRJ2) aircraft were detected at approximately a 95% rate in both frequency bands.

- Small turboprop aircraft generated detectable wake noise; however, the sample size was small and the detection rate was not high.

It is known that every aircraft generates a wake, as a consequence of its generating lift. However, it is not clear from the data whether every wake generates detectable sound. A preliminary examination of the flyby cases corresponding to no-detected-wake-sound included looking at possible influence of the meteorological conditions such as wind magnitude, atmospheric stability classes, and turbulence. However, no correlation was found between meteorological conditions and the absence of a wake acoustic signal.

It was observed that the poor wake noise generators tended to be newer aircraft, presumably having aerodynamically cleaner wings. In the case of some of the Airbus models, a specific physical attribute that likely contributes to low wake noise generation is the relatively simple flap design. Although the "silent vortex" cases represent, in some sense, the most interesting and important cases to study from a sensor development perspective, efforts to date have concentrated on cases that did produce detectable vortex sound. It is also noted that source localization maps for regional jets are typically brighter than those for larger aircraft (using the same color contours), suggesting that the initial stage of the vortices, as imaged acoustically, is not a good indication of wake circulation.

3.2.2 DLR Microphone Arrays

Because the results from the NASA-DOT microphone array measurements were affected by data collection implementation issues such as synchronization uncertainty (Ref. 15), clipping, and non-uniform response of microphones, it is worthwhile to compare the NASA-DOT results to those obtained by the DLR-Berlin group for their microphone arrays.

The DLR-Berlin group operated the two microphone arrays depicted in Figure 3-5 from August 28 to September 4, 2003:

- The more compact "T" array was designed for relatively high frequencies; its data were processed over a 100-1,000 Hz frequency range. Best wake detection was found to be in the 250-630 Hz band.
- The much larger "X" array was designed with the specific goal of higher spatial resolution at frequencies less than 200 Hz. The "X" array data were processed over a 10-400 Hz frequency range; best wake detection was for 40-200 Hz.

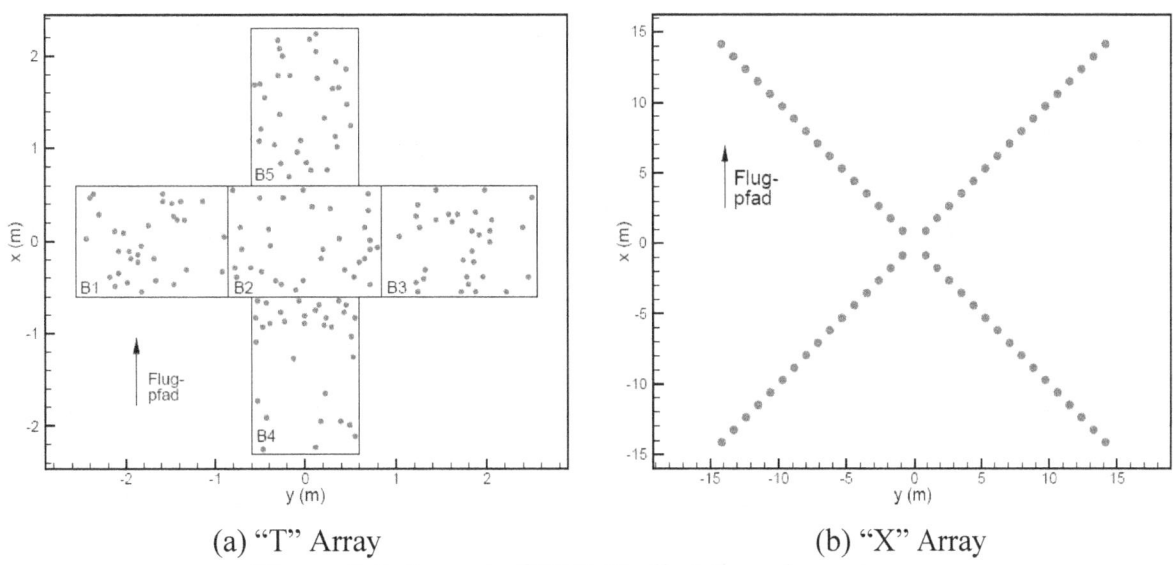

(a) "T" Array   (b) "X" Array

Figure 3-5 Layout of DLR-Berlin Microphone Arrays
Flug Pfad is Flight Path

The outputs of the two arrays were combined to provide altitude tracking information. This was done by considering each of the arrays as a sub-array in a combined split-aperture array configuration. A validation process using the source localization maps in comparison with LIDAR data (similar to that done for the NASA-DOT array) was conducted. Additional details of the DLR setup can be found in Ref. 24 (in German).

Figure 3-6 summarizes the detection statistics by aircraft type for the DLR in the 40-630 Hz frequency band. The gray columns show the fraction of wakes detected at age 10 sec. The other marks are indicative of the wake life, showing the probability of wake signals lasting for 20, 30 and 40 sec. The overall trend as observed by the DLR arrays is similar to that of the NASA-DOT array:

- B-737 and B-757 series aircraft were all detected at 80% probability or greater.
- A-319 (55%) and A-320 (80%) aircraft had lower detection probabilities.
- B-777-200 statistics are similar to those for the NASA-DOT microphones when processed using the 200-400 Hz frequency band.
- BA46 (100%) and CRJ2 (90%), although smaller, had high detection probabilities.

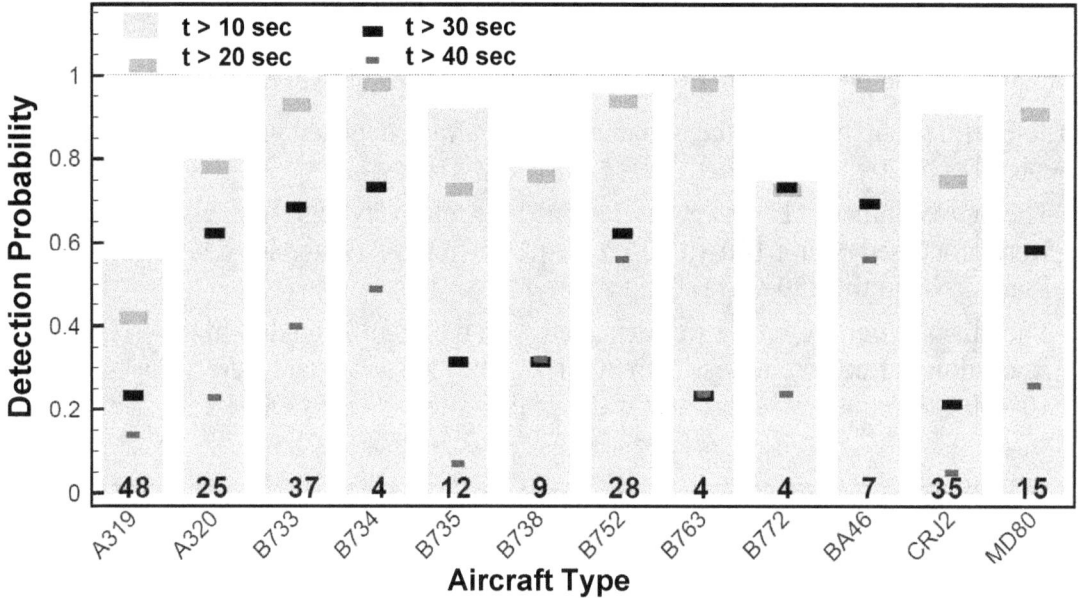

Figure 3-6 Wake Acoustics Detection by Aircraft Type and Duration for the DLR Array
Statistics are for 40-630 Hz frequency band. The number of approaches is listed at the base of each bar.

3.2.3 SOCRATES 2003 Prototype

As stated earlier, contractor testing of the then-latest SOCRATES prototype was an auxiliary objective in the DEN03 test. Due to resource constraints, the Government team focused its attention on investigating wake acoustic phenomenology. Government evaluation of the SOCRATES sensor was limited to correlating its detection and tracking statistics with other wake sensors. The SOCRATES system level evaluation was deferred to the DEN05 test.

During DEN03 a four-beam SOCRATES array was fielded employing 50-m (approximately 164-ft) long laser beams. A beamforming/processing frequency band of 270-360 Hz was used, as was the case in previous SOCRATES testing. With the four-beam array, there was not enough spatial resolution to localize the height and lateral position of a noise source. (This test result is consistent with the theoretical analysis of SOCRATES resolution presented in Section 4.2). Instead, the beamforming results were presented in terms of the elevation angle of a sound source versus time. Figure 3-7 shows a photograph of the partial SOCRATES hardware setup with the definition of the beamforming elevation angle as well as the array response pattern superimposed. The photograph was taken by a person facing southward toward Runway 16L.

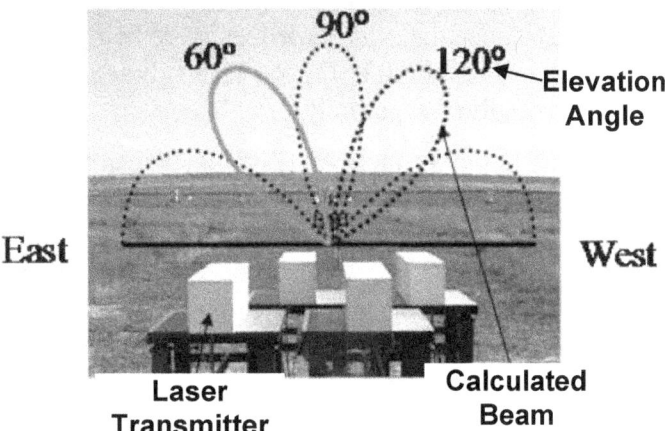

Figure 3-7 Photograph of DEN03 SOCRATES Sub-Array and Calculated Acoustic Beams
Notation used in the SOCRATES angular tracking results is superimposed. Source: FST-LMCo

Results from the SOCRATES prototype sensor were presented by the contractor team in the form of plots of source strength and elevation versus time. An example, for a B-737 flyby (on September 3, 2003, at 08:00 am local time), is shown in Figure 3-8(a). The horizontal axis is the elevation angle, with 0 deg being a sound source to the east on the horizon and 180 deg representing a source to the west. The initial aircraft noise is shown at 90 deg elevation (i.e., over the SOCRATES array) and the time after flyby is indicated on the vertical axis. After the

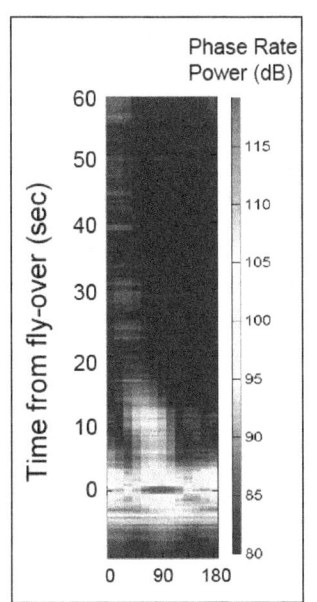 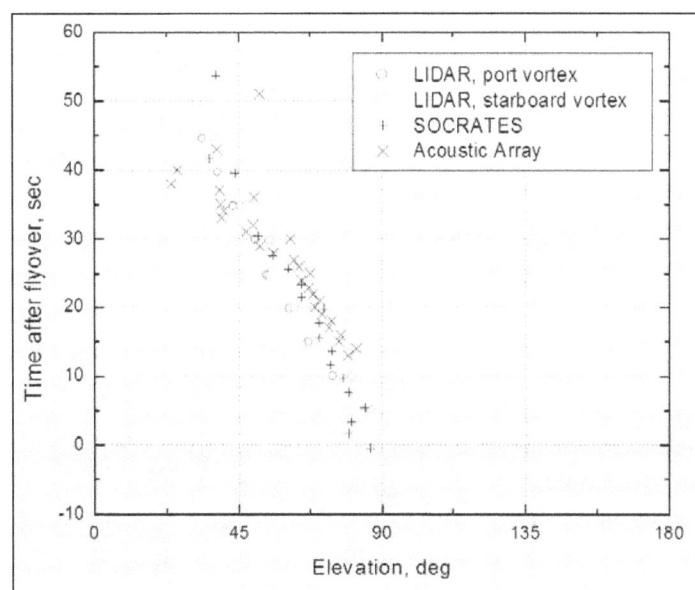

(a) Sample Noise Localization Map (b) Microphone, SOCRATES and Pulsed LIDAR Tracks
Figure 3-8 DEN03 Test Results for Three Sensors

flyby, a fainter sound source is revealed behind the aircraft and drifting eastward.

Figure 3-8(b) is a comparison of the results for the B-737 example shown in Figure 3-8(a). The phased microphone array data for the figure were processed by NASA LaRC and the results, along with Pulsed LIDAR and SOCRATES track data, were superimposed onto the same plot, in the elevation-versus-time format used for SOCRATES (Ref. 21). Results from the three sensors all agree, in that the vortex signature from each sensor drifted eastward. Additional examples of this type were generated and shown in Ref. 21. In the view of the present authors, these provided the first convincing evidence that wake vortices were being tracked by SOCRATES. However, the spatial resolution of the DEN03 SOCRATES sub-array was such that port and starboard vortices could not be resolved separately.

Source localization maps for the DEN03 SOCRATES data such as Figure 3-8(a) were reviewed manually by the FST-LMCo team, and the associated wake detection statistics are shown in Table 3-1, which reports an overall detection rate of 81% for a total of 882 flybys (Ref. 25). The SOCRATES statistics were not independently evaluated by the present authors, and no further attempts were made to quantify the tracking accuracy and wake age relative to the Pulsed LIDAR data. SOCRATES detection statistics are contrasted with those for the microphone array in Section 3.3.

Table 3-1 SOCRATES Detection Statistics for DEN03 Reported by FST-LMCo (Ref. 3)

A-320*	B-737	B-757	B-767	B-777	MD-80	RJ	TOTAL
191 / 245	217 / 283	73 / 92	13 / 15	17 / 17	46 / 54	156 / 176	713 / 882
78%	77%	79%	87%	100%	85%	89%	81%

*The "A-320" category in the FST-LMCo analyses consisted of four Large Airbus aircraft models: A-318, A-319, A-320 and A-321.

3.2.4 SOCRATES 2005 Prototype

Although the 2005 SOCRATES evaluation efforts will be discussed more extensively in Chapter 4 of this report, summary results reported by the FST-LMCo team are provided here. As described in Section 2.2, the DEN05 SOCRATES prototype sensor had 16 beams and experimented with two array configurations, the Skyward-Looking and Billboard arrays (Ref. 25). As was in the case of the 2003 test, one or more technical staff was always present to attend to the SOCRATES instruments, and the measurements took place predominately under VMC.

The Skyward-Looking array was a more extensive version of the DEN03 array. Sub-arrays of eight beams each were situated on both sides of the extended runway center at approximately the same distance from the Runway 16L threshold. The two sub-arrays provided results that were combined and triangulated to provide both lateral and vertical wake tracking.

The DEN05 SOCRATES wake tracks from the Skyward-Looking configuration were compared with Pulsed LIDAR tracks. While the ability to provide range information was a significant improvement over the DEN03 results, the DEN05 array still lacked the spatial resolution needed to distinguish port and starboard vortices. Other important features of the DEN05 array included the use of commercially available optical equipment and more advanced beamforming processing which incorporated techniques to dynamically equalize the output power for all channels ("auto-gain") as well as to suppress RIV noise ("diagonal deletion") — see Ref. 3.

Wake detection statistics from the 2005 Skyward-Looking array configuration compiled by the FST-LMCo team are shown in Table 3-2. For a dataset of 1,987 flybys (more than double the size of DEN03 dataset), the reported overall detection rate was 86%. These results have not been independently verified by researchers outside of the SOCRATES contractor team, but the improved statistic (over the 81% for DEN03) is not entirely unexpected. What is unexpected is the modest increase in the detection rate over the DEN03 test, since four times as many beams were available (16 versus 4) and more advanced processing algorithms were used.

Table 3-2 SOCRATES Detection Statistics for DEN05 Reported by FST-LMCo (Ref. 3)

A-320*	B-737	B-757	B-767	B-777	MD-80	RJ	TOTAL
529 / 697	366 / 404	135 / 144	19 / 20	35 / 40	120 / 128	497 / 552	1,701 / 1,987
76%	91%	94%	95%	88%	94%	90%	86%

*The "A-320" category in the FST-LMCo analyses consisted of four Large Airbus aircraft models: A-318, A-319, A-320 and A-321.

Some aircraft specific variations were observed between the DEN03 and DEN05 results, most notably improved detections for B-737, MD-80, B-757 and B-767. There was also a decrease in the detection statistics for the B-777, but the sample size was comparatively small. A low detection probability for the "A320" aircraft group was again reported.

Statistics for the duration of the DEN05 SOCRATES wake tracks are shown in Figure 3-9. As might be inferred from Figure 3-2, relative to Pulsed LIDAR wake tracks, SOCRATES tracks were significantly shorter in duration. Comparisons of the DEN03 SOCRATES track lengths with those for the microphone array (Subsection 3.3.2) and Pulsed LIDAR (Subsection 3.3.3) are presented below.

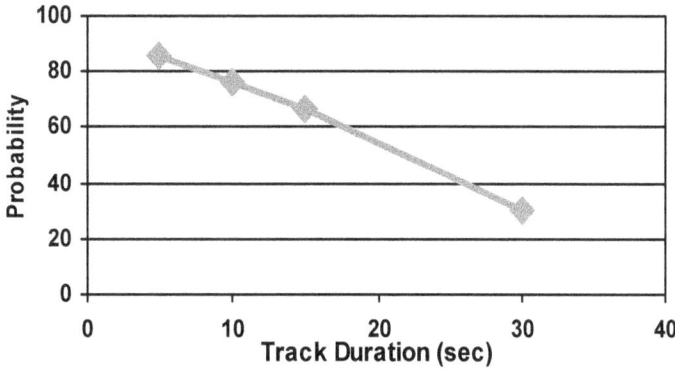

Figure 3-9 DEN05 SOCRATES Prototype Wake Track Duration Statistics (Ref. 3)

The FST-LMCo analysis of the 2005 Skyward-Looking data indicated performance sensitivities to temperature and wind, which are believed to be related to elevated RIV noise. These aspects are more thoroughly discussed in Section 4.3 of this report, as part of the SOCRATES 2005 system evaluation conducted by the Volpe Center.

3.3 SOCRATES 2003 Prototype Comparison to Other Sensors

3.3.1 Wake Detection Comparison (Microphone Array and SOCRATES)

Summary detection statistics reported by the FST-LMCo contractor team for the SOCRATES 2003 prototype have both similarities and differences relative to those for the NASA-DOT microphone array (Figure 3-4). Since the two sensors did not operate during the same time period (see Figure 2-2), their summary statistics were not derived from the same set of arrivals. For comparison purposes, a set of 213 flybys was identified for which the two sensor systems were both collecting data. For this dataset, the microphone array 200-400 Hz beamforming results were used, as this band is a closer match to the SOCRATES processing band.

Results for the SOCRATES-microphone wake detection comparison are shown in Figure 3-10, for both individual aircraft models and the aggregate set of flybys (right-most pair of bars). Overall, the microphone array detected 86% of the 213 approaches, while SOCRATES detected 70% of these tracks. This comparison again revealed the peculiar lower detection statistics of the A-318 through A-321 aircraft series (which were lumped into the "A320" group) for both acoustic sensors. It should be noted that in generating statistics for the two sensors, manual reviews of the source localization maps were conducted. However, no effort was made to coordinate the two scoring schemes. The higher detection statistics for the microphone array may be the result of (a) a broader and lower frequency beamforming frequency range used compared to that for SOCRATES, and/or (b) a possibly more liberal detection threshold used in the visual analysis.

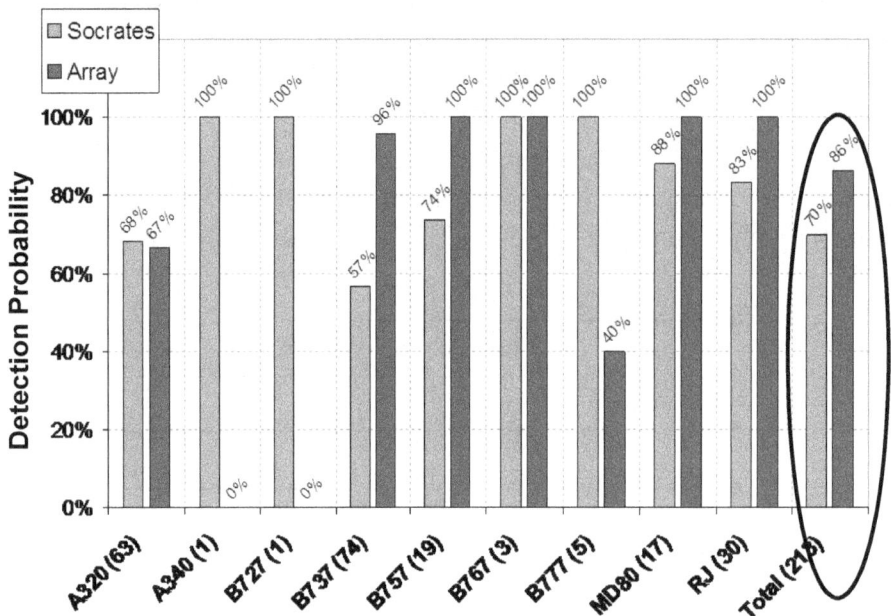

Figure 3-10 Detection Statistics for SOCRATES 2003 Prototype and Microphone Array
Statistics are based on a set of 213 flybys observed by the two sensors.

3.3.2 Wake Track Length Comparison (Microphone Array and SOCRATES)

From the 213 flybys identified when the microphone array and SOCRATES were both operating, there were 84 flybys for which a track length was available for each sensor. The distribution of airplanes for these tracks is shown in Figure 3-11, and the average difference in track length between the microphone array and SOCRATES, by airplane type, is presented in Figure 3-12. For each aircraft type, the average microphone array wake track length was greater than that for SOCRATES, with the difference ranging from 8.4 sec to 29.2 sec. Figure 3-12 suggests that the difference in track length increases with aircraft weight.

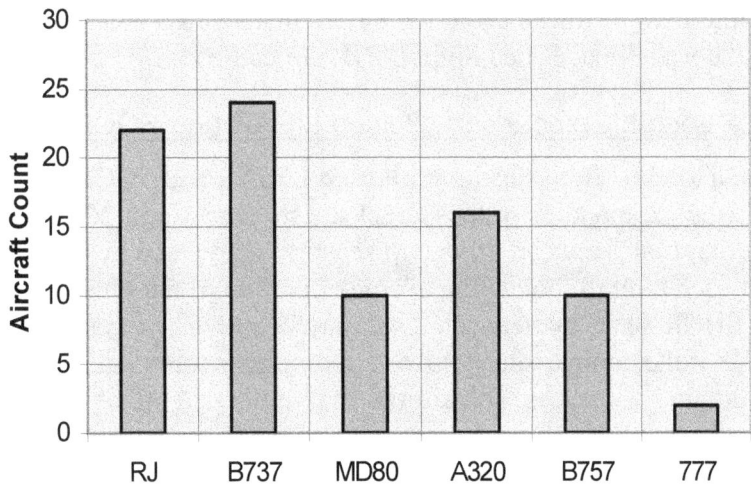

Figure 3-11 Distribution of Aircraft Having Microphone Array and SOCRATES Tracks

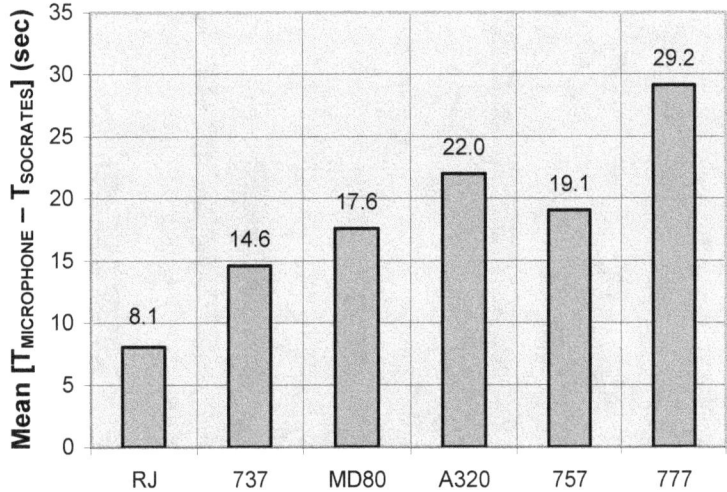

Figure 3-12 Track Length Excess: Microphone Array over SOCRATES 2003 Prototype
Aircraft arranged in order of increasing maximum take-off weight

Calculations of the characteristic acoustic frequency associated with the core frequency of a Kirchhoff vortex (where maximum signal-to-noise ratio [SNR] is expected) revealed that this characteristic frequency decreases with increasing aircraft weight and wing span (Ref. 26). Therefore it is possible that the wake acoustic contribution to higher frequencies also decreases, reducing the SNR in the bands where RIV noise is low. It is however important to note that:

- As in the case of wake detection, the disappearance of a wake from an image is judged by a person observing a sequence of acoustic images. Therefore there is a level of subjectivity in the determination of the length of a wake track.
- Also, as mentioned earlier, the difference in dynamic range and color schemes between the contractor and the Volpe Center presentations of the acoustic images can affect the statistics.
- In this comparison, the microphone array analysis employed the 200-400 Hz band, since this is similar to the band where the SOCRATES sensor operates. It is possible

that longer microphone tracks could have been generated if the 50-200 Hz band was used, since the wake images had higher SNRs in that band.

3.3.3 Wake Track Length Comparison (Pulsed LIDAR and SOCRATES)

For the DEN03 test, a total 67 flybys had both Pulsed LIDAR and SOCRATES tracks. The distribution of aircraft corresponding to these tracks is shown in Figure 3-13. Averages of the differences in track durations between these two sensors, segregated by airplane type, are shown in Figure 3-14. For each aircraft model, the average duration of the Pulsed LIDAR track was longer that that for SOCRATES by between 2 sec and 88 sec. These results are somewhat consistent with those of the comparisons with the microphone array, and generally show that the difference in track duration increases with aircraft maximum takeoff weight.

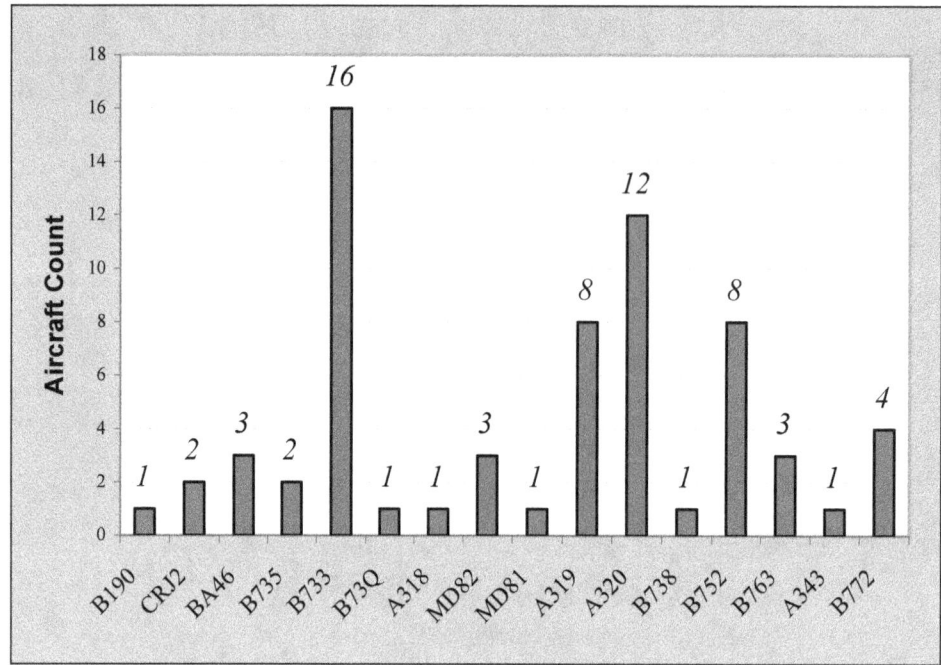

Figure 3-13 Distribution of Aircraft Having both Pulsed LIDAR and SOCRATES Tracks

When comparing the Pulsed LIDAR and SOCRATES tracks, the following should be taken into account:

- The two sensors did not have a common track termination criterion: SOCRATES terminated when the array gain was less than 5 dB, while the LIDAR terminated based on a likelihood ratio model reaching a circulation threshold of 55 m^2/sec.
- SOCRATES was not able to resolve two vortices, so the calculated age was that of either the wake oval or the longer lasting individual vortex. However, the LIDAR resolved both vortices, and the calculated age was the average of the two. This tended to bias the LIDAR wake age low and the SOCRATES age high.

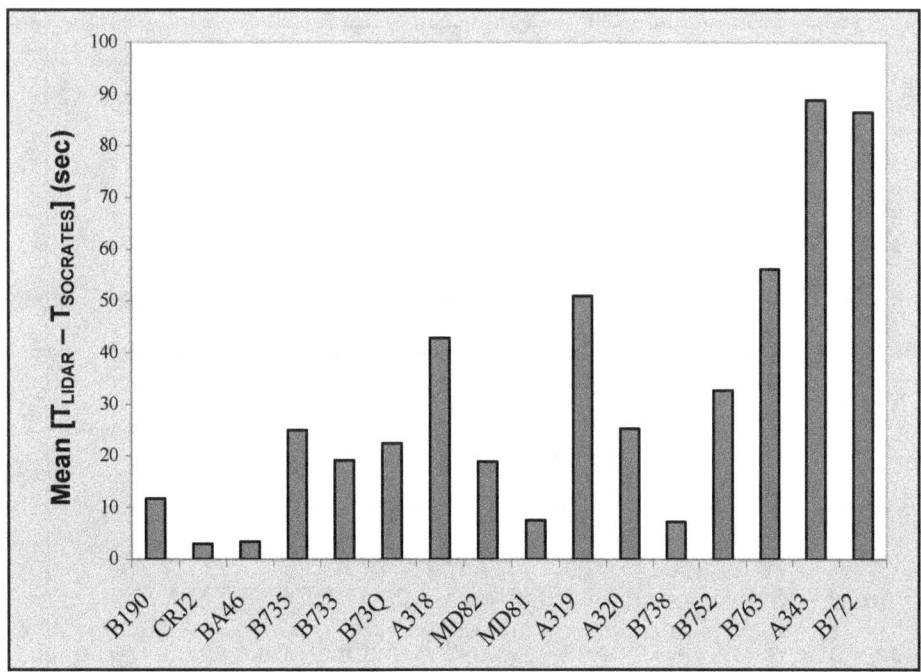

Figure 3-14 Track Length Excess: Pulsed LIDAR over SOCRATES 2003 Prototype
Airplanes arranged in order of increasing maximum take-off weight.

3.3.4 Effect of Runway Crosswind on Detection

Since the performance of a wake sensor is better understood when examined with respect to airport operational parameters and operational modes, the DEN03 SOCRATES detection statistics were correlated with the crosswind derived from the Pulsed LIDAR, which scanned over the microphone array and 400 ft from the SOCRATES prototype sensor. The crosswind profile from the LIDAR was averaged from 6.5 ft to 975 ft above the phased microphone array. The resulting crosswind distribution is presented in Figure 3-15, with positive winds being from the west.

Overall, it was found that DEN03 SOCRATES tracks were not available when the crosswind was more than 15 kt, even though landings were conducted with greater than 15 kt of crosswind during the period of SOCRATES operation. A similar type of crosswind limit was also observed in the phased microphone array data. This aspect of the data analysis was repeated (with more operations) for the DEN05 data and the Skyward-Looking array, and a decrease in SOCRATES wake detection probability when the wind increased was also found by the FST-LMCo team (Ref. 3). RIV noise arising from ambient turbulence passing the SOCRATES laser beam is suspected; this aspect was further studied by the Volpe Center and is presented in Chapter 4 of this report.

3.4 Frequency of Most Consistent Vortex Sound

After establishing the existence of wake acoustic signals as well as characterizing their detection statistics for different sensors, the question of the most consistent frequency range of those signals was examined using microphone array data. For this analysis, spectra of the wake vortices sound and the ambient background noise were generated from the vertical beamforming process (Ref. 18) using a 2-sec processing interval.

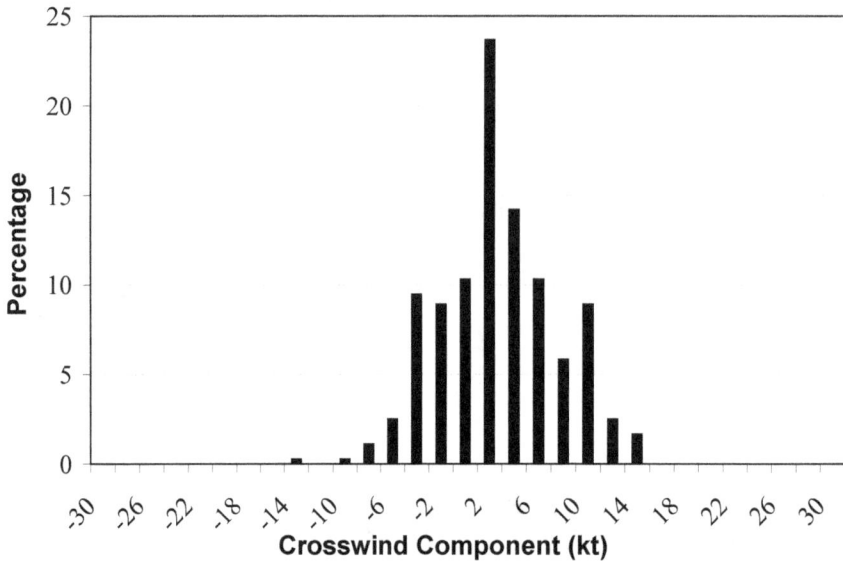

Figure 3-15 Crosswind Distribution for the DEN03 SOCRATES Dataset

In the subsequent discussion, the concept of non-dimensional wake age is introduced. In this analysis, t=0 corresponds to the time when the aircraft flew over the array. The subsequent wake age relative to the flyby time is then normalized by the theoretical time it takes for a vortex pair to sink by 79 percent (more specifically, by $\pi/4$) of the wingspan. Use of non-dimensional time allows the physical wake ages from a diverse group of aircraft to be compared. In general, wake dynamics within the first non-dimensional time interval are considered to be insensitive to atmospheric conditions.

Results of the beamformed spectra were averaged across integer non-dimensional time intervals for the wake (i.e., 0 to 1, 1 to 2, etc.), and the frequency corresponding to the largest energy excess of the wake above the ambient noise was then identified. Figure 3-16 shows an example, taken from a B-737 flyby, where the wake and ambient noise spectra are indicated respectively as green and blue lines. For this flyby, the wake energy that is consistently above the background is approximately between 10 and 100 Hz, with peaks in the 52-60 Hz range throughout the first three non-dimensional time intervals. The durations of one non-dimensional time interval for the two aircraft models considered (B-737 and B-757, both of which are abundant in the database) are given in Table 3-3.

Histograms of the frequency corresponding to the largest energy excess above the ambient background for the B-737 and B-757 are shown in Figure 3-17. These statistics encompass all meteorological conditions encountered during the DEN03 test. The wake acoustic frequency range that was most consistently above the ambient noise for the two aircraft was almost entirely less 100 Hz. It is also noted that the peak of the histogram decreases from B-737 to B-757. The general value of this peak frequency and its scaling with aircraft size is consistent with the rotational frequency of the vortex core (Ref. 26).

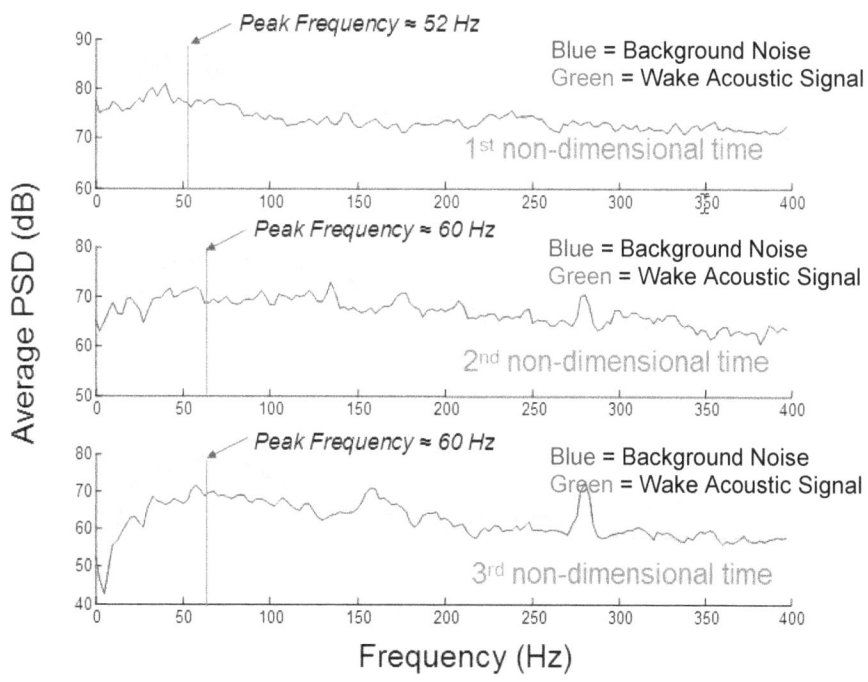

Figure 3-16 B-737 Wake and Background Spectra for Non-Dimensional Time Intervals
The largest energy excess of the wake is indicated by the vertical red line.

Table 3-3 Duration of a Non-Dimensional Time Interval for B-737 and B-757

Aircraft Type	Non-Dimensional Time Interval
B-737	11.5 sec
B-757	16.2 sec

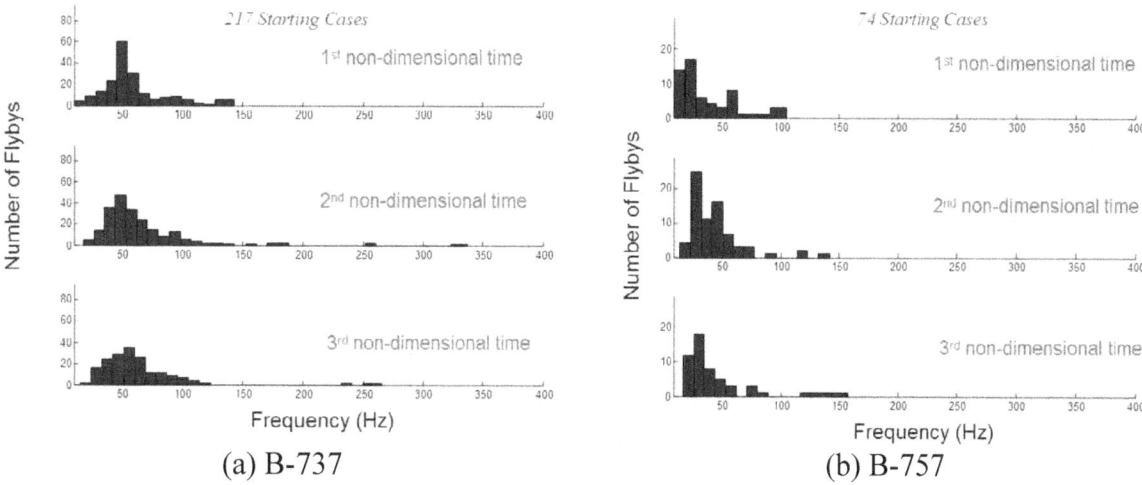

(a) B-737   (b) B-757

Figure 3-17 Histogram of the Largest Energy Excess for B-737 and B-757
The most consistent portion of the vortex sound is approximately 0-100 Hz.

3.5 Relationship of Wake Acoustics and Wake Hazard

3.5.1 Estimating Wake Circulation from Acoustic Power

Based on the vertical sink rate for vortex images derived from the microphone array and the spacing of the imaged vortex pairs, circulation (i.e., the strength of the flow field) was estimated for the first non-dimensional age interval, where the influence of weather is less important (Ref. 16). Such an inferred circulation has been useful as a consistency check on several sensors.

However, for the purpose of quantifying the potential wake hazard level in the context of separation standards, it is the older portion of the vortex that is of most interest. To estimate the strength of older wakes, data from the microphone array were analyzed as follows. Using the Kirchhoff vortex model as a guide (Ref. 26), supplemented by the acoustic energy peak distribution data shown in Figure 3-17, acoustic power above ambient background around the peak frequency was obtained and plotted as a function of time. The choice of the frequency band was motivated by the fact that vortex core dynamics represents a mechanism that is always present and its associated acoustic signal is believed to be more consistent than the signal generated by turbulence around the core. Since a power law relationship between acoustic power and velocity is found in many aeroacoustic phenomena (Ref. 27), Pulsed LIDAR circulation data corresponding to the microphone array tracks were also extracted for comparison.

It was found, as expected, that on average both acoustic signal power and LIDAR circulation decreased over time — e.g., Figure 3-18 (B-737, involving 231 flybys) and Figure 3-19 (B-757, involving 76 flybys). However, when the data were cross-correlated, only a weak correlation coefficient was found, and the acoustic power time histories were significantly shorter (approximately 60%) in duration than the Pulsed LIDAR circulation histories. Therefore, this study does not support the possibility of reliably inferring a hazard level from the power or intensity of the wake acoustic emissions. The observation that regional jets often represent some of the loudest wakes further suggest the difficulty of relying on vortex sound level to infer wake hazard using acoustic power or intensity alone.

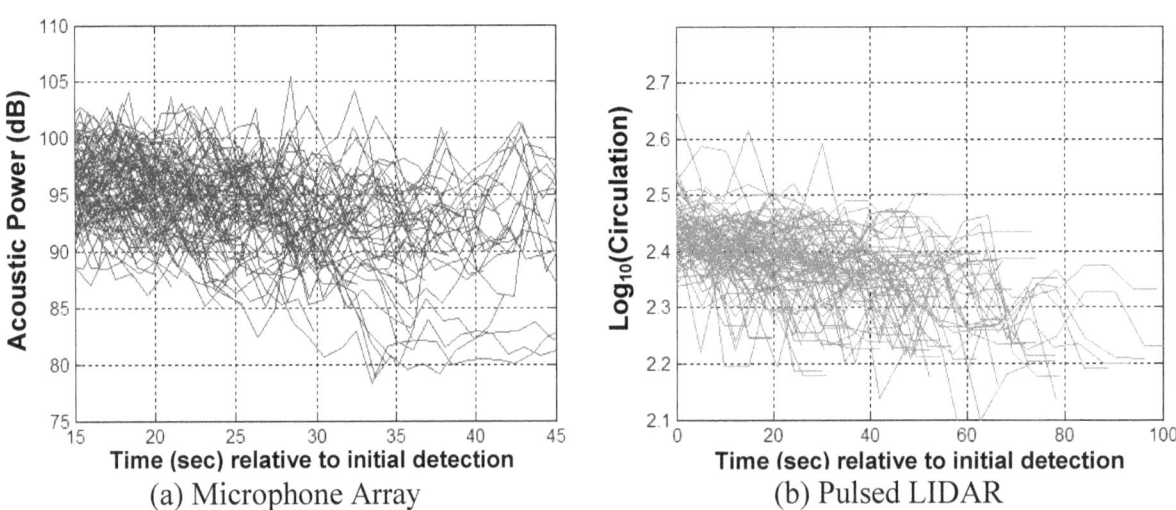

(a) Microphone Array   (b) Pulsed LIDAR

Figure 3-18 Comparison of Wake Acoustic Power and LIDAR Circulation for B-737

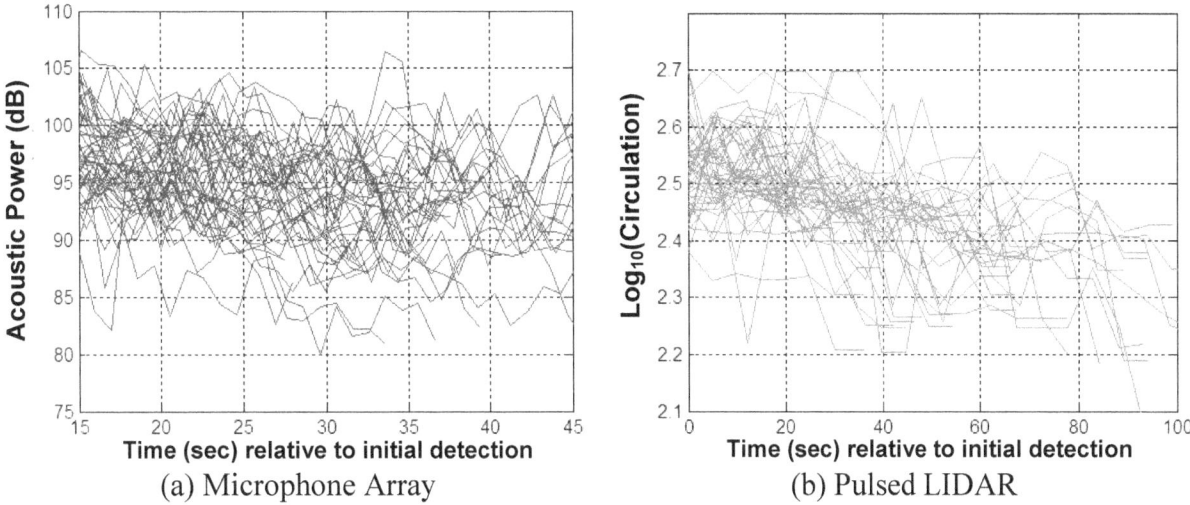

(a) Microphone Array (b) Pulsed LIDAR

Figure 3-19 Comparison of Wake Acoustic Power and LIDAR Circulation for B-757

3.5.2 Possible Role of Acoustics in Estimating Wake Axial Coherence

There exists a second scheme for identifying a greatly reduced wake hazard level. It has long been argued that circulation of the wake is an important, but not the only, parameter for characterizing wake vortex hazard. The vortex axial coherence (i.e., how curved vortices are) during an encounter, as well as the control authority of the following aircraft, also play important roles (Ref. 28). For example, a recent flight simulation study suggested that encounter with vortices in Crow instability or other highly contorted states (Ref. 29) greatly reduces the maximum bank angle experienced by the penetrating aircraft. If flight research confirms these findings, and if contorted vortices are visualized though source localization maps in a passive wake acoustic sensor, a qualitative indication of less hazardous wakes is then potentially feasible. In that case, the relatively poor spatial resolution of SOCRATES, relative to the microphone array, would be a concern.

4. 2005 SOCRATES Sensor Characterization

The SOCRATES concept is based on two premises: (1) each airplane's wake vortices generate acoustic energy that a properly designed sensor on the ground can "hear" in the presence of other noise sources throughout their duration as a hazard to other aircraft; and (2) a ground-based SOCRATES opto-acoustic sensor can detect wake sound sufficiently well to be valuable for research and/or operational purposes. Premise (1) was addressed in Chapter 1, where analyses of aircraft wake detectability and tracking by microphone arrays and SOCRATES are presented.

This chapter addresses premise (2) by analyzing the performance of the DEN05 prototype SOCRATES sensor as well as potential extensions of that sensor (e.g., involving additional beams or different noise levels). Section 4.1 first describes the DEN05 SOCRATES system. Section 4.2 then analyzes its performance in terms of its ability to resolve wake vortices. Section 4.3 analyzes the effect of noise sources on the DEN05 SOCRATES prototype. In particular, an analysis of Refractive Index Variability (RIV), which is judged to be the dominant SOCRATES noise source up to frequencies greater than 200 Hz, is presented. Finally, Section 4.4 discusses two other SOCRATES performance issues: near field effects that can limit the benefits of using longer beams, and atmospheric effects that can limit benefits of both longer and an increased number of beams.

4.1 Description

4.1.1 Principle of Operation

When an optical beam propagates through air, spatial variations of the refractive index along its path impose spatially varying phase shifts on the beam's wavefronts. This results in a cumulative phase shift of the detected optical beam. If an acoustic wavefront (which has an associated acoustic refractive index variation of its own) passes through the optical beam, the acoustic pressure fluctuations modulate the along-beam optical refractive index profile, resulting in time variation of the detected optical beam phase. If the phase of the optical beam can be accurately measured, it is conceivable that the beam can act as a microphone. For a distant acoustic source, the maximum sensitivity of this optical microphone is realized when the normal to the acoustic wavefront is perpendicular to the optical beam's path (center of Figure 4-1) and degrades as the angle between the normal to the acoustic wavefront and optical beams decreases from 90 deg.

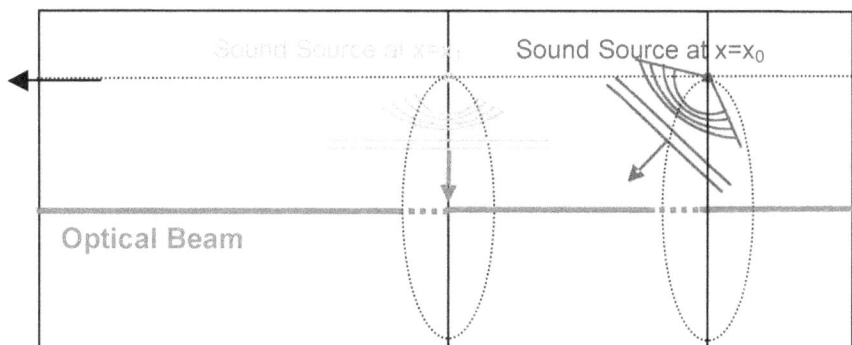

Figure 4-1 Illustration of Acoustic Wavefronts Impinging on Optical Beam

If a nearby omni-directional noise source moves in the longitudinal direction (parallel to the beam, along the dotted line in Figure 4-1), it will induce the maximum response when it is in the perpendicular plane passing through the center of that beam. The response decreases as the source moves away from this plane. However, the beam response is insensitive to lateral motion of the source (perpendicular to the longitudinal direction). For example, as one of the sources shown in Figure 4-1 traces the corresponding dotted circle, the beam response would not change.

When multiple parallel beams are deployed and their responses are processed coherently, angular selectivity and tracking in the lateral direction is possible. SOCRATES uses coherent processing to perform angular tracking of the lateral motion of wakes.

4.1.2 Deployed Configurations

During the DEN05 campaign, a prototype SOCRATES array was deployed in two different configurations. The chronologically first deployed, and the most tested, configuration — termed Skyward-Looking — involved a total of 16 100-m (approximately 328-ft) long beams deployed at the same height (3.5 ft) above the ground, with a common lateral spacing of 1.5 ft between adjacent beams. The Skyward-Looking configuration (Figure 4-2) consisted of two eight-beam sub-arrays separated by approximately 200 m (656 ft) oriented parallel to the runway extended centerline.

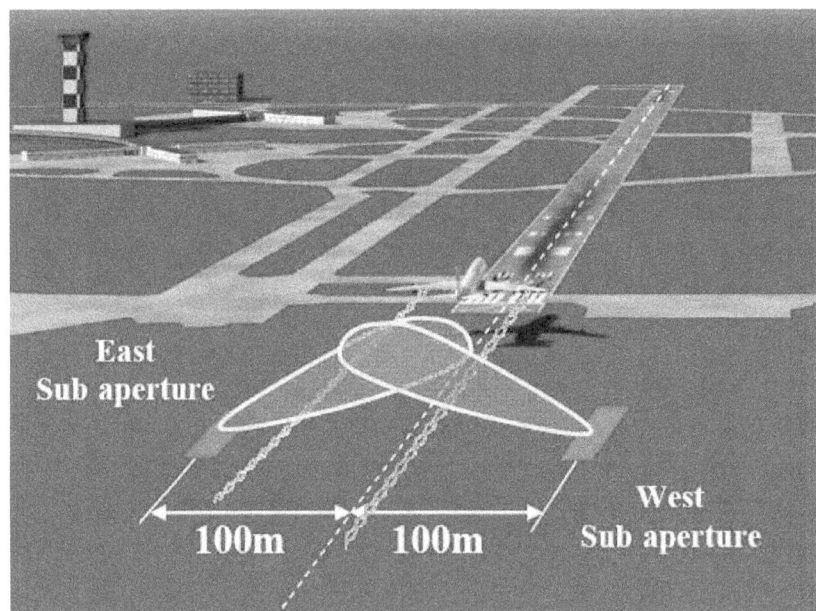

Figure 4-2 DEN05 SOCRATES Skyward-Looking Configuration (Conceptual, Ref. 3)

The 1.5-ft beam spacing within each sub-array corresponded to one-half an acoustic wavelength at 360 Hz frequency, the highest frequency used in processing Skyward-Looking array data, and was presumably selected (as is standard in uniform array design) to ensure that the elevation angle of a sound source could be unambiguously determined. Having elevation angle measurements from two sub-arrays enabled vertical and lateral detection and tracking of wake locations. The DEN05 sensor, with more beams and improved data processing, provided better angular resolution than the DEN03 prototype (discussed further in Section 4.2).

The Billboard array (also called the Range-Focused array, Figure 4-3) was the second configuration tested. It consisted of three identical panels ("billboards"). Each panel had five 100-m (approximately 328-ft) long parallel beams, stacked vertically with a separation of about 4 ft between adjacent beams. The three panels were not placed in a straight line; instead there was a slight angular offset (approximately 5.8 deg) between the bases of adjacent panels, in order to point the beams toward a common focal point.

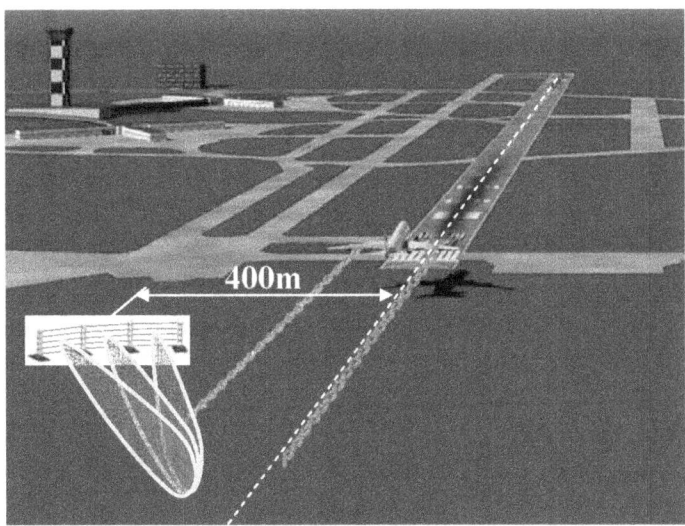

Figure 4-3 DEN05 SOCRATES Billboard Configuration (Conceptual, Ref. 3)

First tried in the DEN05 test, the Billboard configuration was designed for a standoff listening concept, with the aircraft track being approximately 1,000 m (3,280 ft) from the array. The 4-ft beam separation corresponded to maximum frequency of 140 Hz for unambiguous determination of the sound source elevation angle. Apparently the lower maximum processing frequency was selected to counteract the greater signal attenuation expected over the longer propagation path (roughly five times that for the Skyward-Looking array).

Results for the Billboard configuration have not been presented to the Government. Reference 3 states:

> "Post-test analysis of the range-focused array data provided proof of the range-focus concept. However, coherent processing of the signals from the individual billboard "panels" was not successful and incoherent processing resulted in only limited detection of wake vortices. Several questions must be answered before this array design can be used in practice. For example, it is hypothesized that the wake vortex signal has a directivity pattern with limited acoustic energy emitted in directions close to the axis of the vortices, and this directivity pattern precluded the range-focused array from hearing the vortex sound. Also, it is possible that the wake acoustic signal is sometimes masked by the effect of nonlinear optical demodulation "smearing" the low frequency tower vibration energy to the higher frequencies. Both of these issues are under investigation."

Therefore, the analysis from this point forward is limited to the Skyward-Looking array.

4.1.3 Hardware Components

A simplified block diagram of the DEN05 SOCRATES sensor electronics is shown in Figure 4-4. Energy from a 1,319 nm (1.319 μm) CW laser was coupled into a single-mode fiber and phase modulated using a piezo-electric modulator. The modulation wave form was sinusoidal with a drive frequency of 11,038 Hz. The energy then was fed into a 1-to-32 beam-splitter and 16 channels went to the transceiver heads outside of the electronics enclosure (the other 16 channels were not used in the DEN05 prototype). At the fiber-air interface of each Transmit/Receive (T/R) head, 5% of the power was back-reflected to provide a local oscillator. The remaining 95% of the power left the end of the fiber and was coupled into the air by appropriate focusing optics, forming a sensing beam.

Each beam traveled in free space, reflected back from a cat's eye[*] and returned. There, through the focusing optics, it was coupled back into the fiber. The returning energy was mixed with the 5% of back-reflected light before it entered the optical isolator and the result was coupled into the receive channel (Figure 4-5). The returned beam was then sensed by the Optiphase optical receiver and the output signal was digitized at a rate of 12 samples/modulation cycle. The optical isolator in the T/R head ensured that the mixed optical signal did not propagate back into the splitter. The digitized signal was then processed by the Optiphase digital demodulator (Refs. 30 and 31) to recover the phase shift of the sensing beam relative to the local oscillator. A 32-bit counter was used to keep track of the integrated phase up to $\pm 20\pi$. This phase was subsequently unwrapped[†] and stored as raw data for each of the beams.

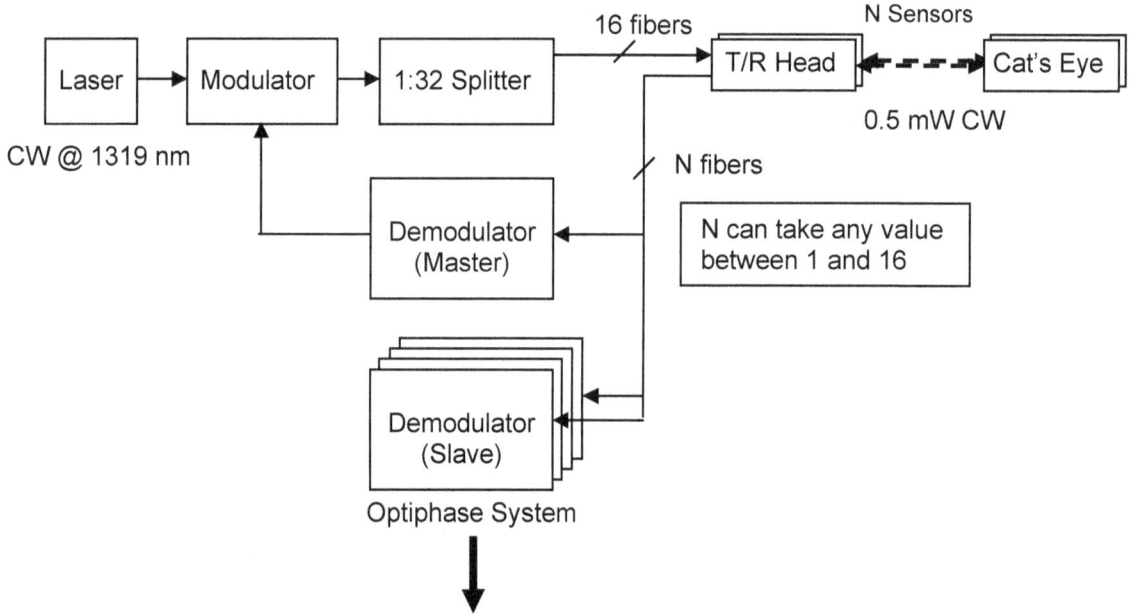

Figure 4-4 Simplified SOCRATES DEN05 Block Diagram

Major differences between the DEN03 and DEN05 prototype system hardware were:

---

[*] A cat's-eye is a retro-reflector that effectively re-directs the laser beam back toward the source region.
[†] Phase unwrapping is the conversion of a stream of phase data from a cyclical form (e.g., in the range $\pm 2\pi$) to a "continuous" form that does not have large discontinuities.

- The 2003 system was pulsed. That is, the laser energy passed through an acousto-optic modulator that acted as an on/off switch and a frequency shifter;
- The 2003 system's demodulation was analog; and
- The output of the 2003 system was phase rate (as opposed to phase).

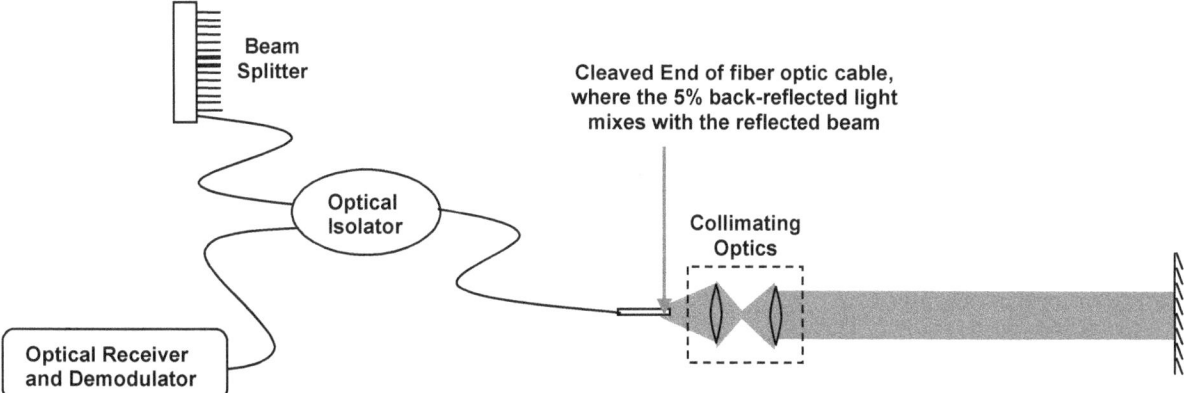

Figure 4-5 Optical Diagram for a Single Channel of the SOCRATES System

However, the discussions in the remainder of this chapter are generally also relevant to the DEN03 SOCRATES system, as the basic principles of the two prototype sensors were similar.

4.2 Array Beampattern

As described in Section 4.1, each laser beam inherently acted like a spatial filter in the longitudinal (along-beam) direction. Based on information provided by FST, the Volpe Center performed simulations to predict the degree of filtering for an eight-beam array 100 m (approximately 328 ft) in length having 1.5 ft beam spacing. Figure 4-6 shows the calculated beam response pattern in the longitudinal direction for a monopole noise source at two heights

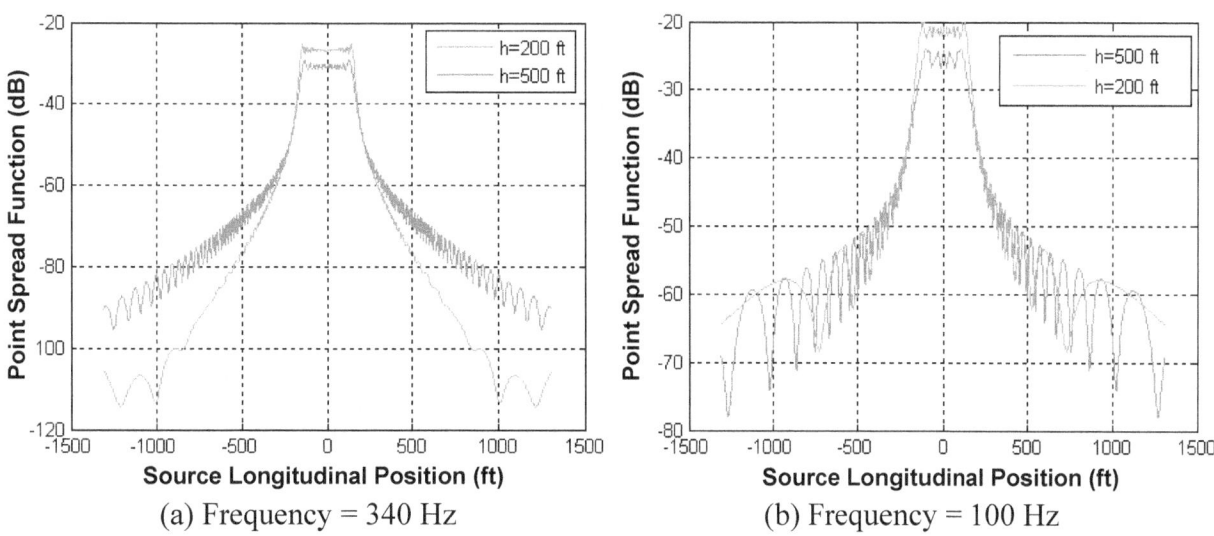

(a) Frequency = 340 Hz (b) Frequency = 100 Hz

Figure 4-6 Predicted SOCRATES 100-m Laser Beam Longitudinal Response
Monopole acoustic source assumed.

(200 ft and 500 ft) of interest for descending wakes and two frequencies (100 Hz and 340 Hz) that approximately bound the band used for processing Skyward-Looking array data. These calculations predict approximately a 30 dB (factor of 1,000) reduction for 100 Hz, and a 40 dB (factor of 10,000) reduction for 340 Hz, in received acoustic power when an approaching or receding aircraft was 500 ft or more away from the arrays relative to being directly above it. This ensured a high degree of confidence that, after an aircraft has passed, the acoustic energy processed by the SOCRATES Skyward-Looking configuration originated from directly above the array rather than from the aircraft location.

Spatial filtering in the lateral direction (perpendicular to the direction of the beams in the horizontal plane) was accomplished by coherent processing (beamforming) each sub-array separately. Figure 4-7 shows the predicted response in the lateral direction of the acoustic beam pointed directly upward, for the same heights and frequencies used for Figure 4-6. As expected from the physical shape of the array (longer dimension was parallel to the runway extended centerline), sidelobe suppression is much more effective in the longitudinal direction than the lateral. In either direction, the sidelobe suppression worsens with decreasing frequency and increasing source height.

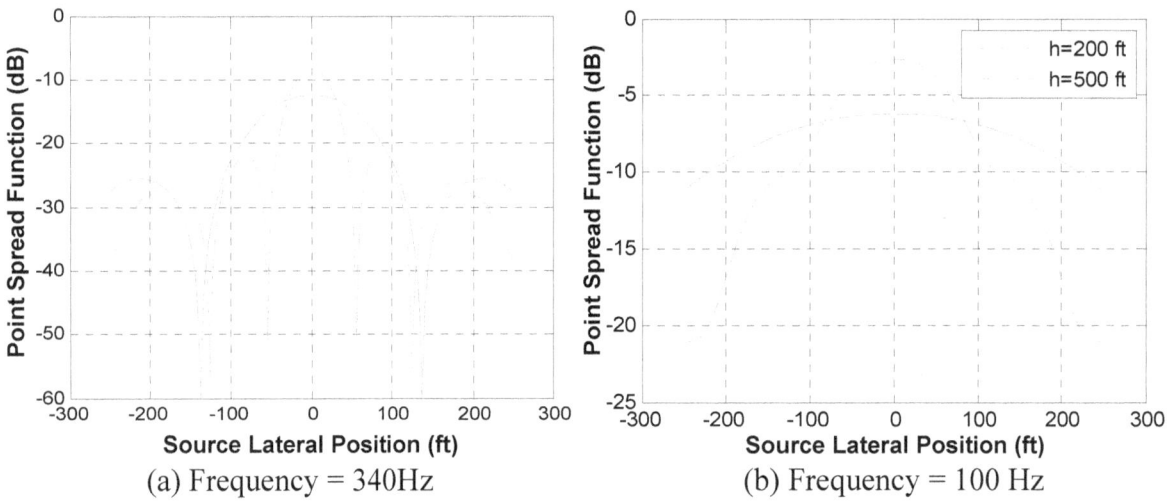

(a) Frequency = 340Hz (b) Frequency = 100 Hz

Figure 4-7 Predicted SOCRATES DEN05 Array Vertical Beam Transverse Response
Monopole acoustic source assumed (PSF is in dB).

Resolution in the lateral direction was best for a source directly above the array, and degraded approximately as the reciprocal of the cosine of the angle from vertical as the coherent processing steered the array. The physical explanation is that the array "appears" narrower to a laterally-displaced acoustic source.

Table 4-1 gives the predicted 3-dB resolution for the response curves shown in Figure 4-7. The initial spacing of Large category aircraft wake vortices is approximately 80 ft (e.g., B-737-600), so that a resolution of roughly 20 ft would be needed to resolve individual vortices. From these calculations it can be predicted that the eight-beam array lateral resolution would not have been sufficient to resolve two vortices at these altitudes. Instead, the array detected the "wake oval" encompassing the two vortices. This prediction is consistent with results of data processing performed by the FST-LMCo team and presented in Subsections 3.2.3 and 3.2.4.

Table 4-1 Predicted DEN05 SOCRATES 8-Beam Sub-Array Transverse 3-dB Resolution

Source Height \ Source Frequency	100 Hz	340 Hz
200 ft	145 ft	47 ft
500 ft	400 ft	115 ft

If additional laser beams could be deployed, the number of beams[*] needed to achieve a resolution of 20 ft directly above the sub-array with 1.5 ft uniform spacing is shown in Table 4-2 Depending upon the wake height and the lowest processing frequency of interest (which depends on the ability to mitigate the effect of RIV noise), implementing such a sub-array would require deploying between 20 beams (for the lower altitude and processing restricted to frequencies near 340 Hz) and 160 beams (for the higher altitude and processing over a 100-340 Hz band, which would likely yield higher detection probabilities). To measure wake heights, two sub-arrays are needed. Thus, a SOCRATES system that achieves a 20 ft lateral resolution and determines wake height would need approximately twice the number of beams shown in Table 4-2.

Table 4-2 Predicted SOCRATES Sub-Array Number of Beams Needed for 20-ft 3-dB Transverse Acoustic Beam Width

Source Height \ Lowest Source Frequency	100 Hz	340 Hz
200 ft	64 beams	20 beams
500 ft	160 beams	48 beams

This beampattern analysis does not account for the fact that resolution is degraded when the source is not directly above the array. It also does not account for the effects of noise sources or lack of coherence of the wavefronts impinging on the laser beams (which are the subject of the next two sections). Thus, this analysis may be optimistic in terms of the benefits of adding beams to a SOCRATES sub-array.

4.3 Effect of RIV Noise

4.3.1 Internal and Environmental Noise Sources

The following possible SOCRATES system noise sources have been identified:

- **Phase noise:** Because back-reflected light from the fiber-air interface was used as the reference beam, without appropriately delaying it to match the optical path length of the sensing beam, the phase of the laser was required to have little variation over the travel time of the sensing beam relative to the acoustically induced phase variations. This requirement also applied to modulator phase noise.

---

[*] The number of beams calculated here is based on design with a uniform spacing of 1.5 ft between beams; it might be possible to reduce the number of beams by employing a non uniform array design.

- **Scintillation Noise:** The digital demodulation technique used by Optiphase required that changes in power of the received sensing beam (e.g., caused by attenuation variations) be negligible over a modulation cycle of approximately 90 μsec.
- **Polarization mismatch between sensing and reference beams:** A mismatch in the polarization of these two beams over a modulation cycle can manifest itself as a phase shift at the demodulator output. If this mismatch was small and slowly varying relative to the time of the modulation cycle, then its effect on the output could be neglected.
- **RIV Noise:** Atmospheric turbulence could induce changes in the refractive index that were significant relative to changes caused by the acoustic signal of interest, and are believed to have been the dominant noise source up to at least 200 Hz.

The remainder of this section addresses the effects of RIV noise on the SOCRATES opto-acoustic sensor. First, the experiment is described during which needed atmospheric turbulence data were collected (Subsection 4.3.2). The RIV simulation is then described (Subsection 4.3.3). Finally, simulation results are presented (Subsection 4.3.4).

4.3.2 Atmospheric Turbulence Data Collection

To quantify atmospheric turbulence on the scales needed for this investigation, high frequency temperature data were collected at DEN on December 16, 2005 (Figure 2-17, Figure 2-18, and Figure 4-8). During the turbulence experiment, fine-wire thermocouples (Campbell Scientific model FW05) were used for *in-situ* calibration of Cold-Wire (CW) anemometers (TSI model 1276-P.5) which provided temperature data for frequencies up to 300 Hz. Each of the CW channels was appropriately amplified and low-pass filtered before being sampled at a rate of 768 Hz. After the *in-situ* calibration, average Power Spectral Densities (PSDs) for all CW anemometers were calculated every 3 min, to provide a spatially-averaged time-varying PSD input to the RIV noise simulation. Sonic and hotwire anemometer data were also collected for the purpose of measuring the turbulent flow field and studying its correlation with RIV noise.

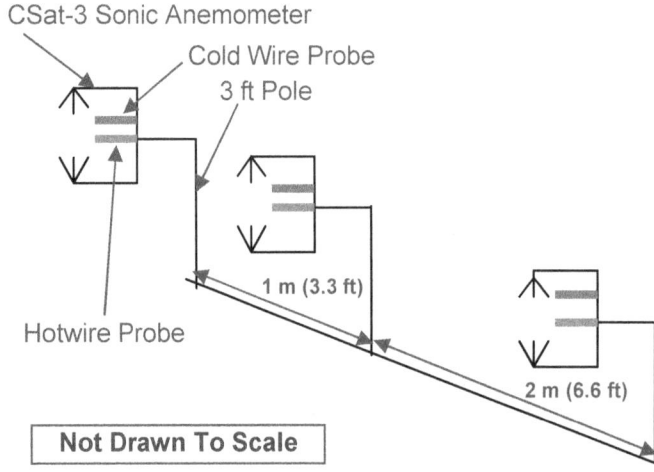

Figure 4-8 Setup for Collection of Near-Ground Turbulence Data

4.3.3 RIV Noise and Acoustic Signal Simulation Descriptions

The Volpe Center developed and exercised simulations of a SOCRATES laser beam's response to a wake acoustic source in the presence of RIV noise, in order to quantify the effects of RIV noise on the sensor's ability to detect and track vortices in the frequency band encompassing the majority of wake acoustic energy. DEN03 microphone array measurements, along with other efforts, indicated that the bulk of wake acoustic energy lies in the 0-100 Hz band for both B-737 and B-757 aircraft (Section 3.4 and Figure 3-17). Moreover, modeling based on sound generation by the unsteady vorticity of the vortex core pointed to frequencies below 100 Hz as being dominant for almost all Large and Heavy aircraft models (Ref. 26).

RIV Noise Simulation — The RIV simulation accepted as input a time-varying temperature PSD, and simulated the index of refraction variability along the length of the beam. This was done by applying the Gladstone-Dale law (Ref. 32) for temperature fluctuations to approximate index-of-refraction fluctuation at measurement points along the laser beam. Since only three measurement points were available, it was not possible to directly create a time series of the index of refraction profile across the beam. Instead, the average of the three PSDs was used to estimate the refractive index PSD over the analysis time interval (T_A=3 sec) in one sub-section of the beam; this PSD was then assumed to pertain along the entire beam.

Beam response to the refractive index temporal and spatial fluctuations was simulated using a Monte-Carlo approach to generate realizations (time series) of the beam noise. In order for this 3-sec PSD to better represent the average 3-sec PSD along the length of the beam, 60 consecutive PSDs were averaged (i.e., over 3 min). The 3-min averaging time was chosen because, even for a low wind speed of 2.5 kt (approximately 1.25 m/sec), 3 min is sufficient for a volume of air with along-wind length greater than two times the beam length to pass through each temperature sensor.

Figure 4-9(a) is an overlay of the successive PSDs of the simulated RIV noise using 5 hr, 48 min of collected temperature data as input. Figure 4-9(b) shows the time variation of the simulated in-band RIV noise over the same time span. The time origin corresponds to about 1:30 pm local Denver time. The RIV noise level was highest in the afternoon and dropped on the order of 8 dB after sunset. This behavior was consistent with expectations, because up to mid afternoon the ground is a heat source, which fuels buoyancy convection near the surface. These structures generate, through hydrodynamic instabilities, smaller structures with a wide range of scales. Towards sunset the ground starts to cool, rendering the near-ground atmosphere more stable. Therefore, if sunset is not accompanied with a significant increase in wind speed, the RIV noise will generally decrease. This effect is expected to be much more prominent in the summer, when warming and cooling of the ground surface usually span a much wider temperature range.

Acoustic Signal Simulation — The wake acoustic signal was modeled as a line of coherent point sources in the 80-100 Hz band. The bulk and most consistent part of the wake energy was found to be in the first 100 Hz. The upper part of this band was chosen (80-100 Hz) as a best-case scenario, since the RIV noise drops with frequency.

The line source representing the wake was assumed to be parallel to the optical sensing beams at 900 ft AGL. The simulated wake length was taken to be equal to that of the optical beam, since contributions from sources in the rest of the wake will be rejected by the beam's spatial selectivity pattern. A 2-ft core size was used (consistent with the predicted core diameter of a B-737), and the eccentricity used was e_{cc}=0.03. According to Ref. 26, the wake acoustic source strength in the core rotation frequency band is a strong function of eccentricity. Therefore a relatively large value of eccentricity was used to simulate a loud wake (source strength equal to 130 dB-SPL 1 m from a monopole source). The 130 dB of power was distributed evenly over the line source and frequency band.

A free-space Green's function was used to propagate the source acoustic waves to each point of the beam, in order to calculate the spatial and temporal variations of the acoustically-induced refractive index along the beam's path. The beam response was calculated and added to the simulated noise. PSDs of the simulated total beam response were calculated for each analysis time interval T_A in order to quantify the in-band SNR variation over time.

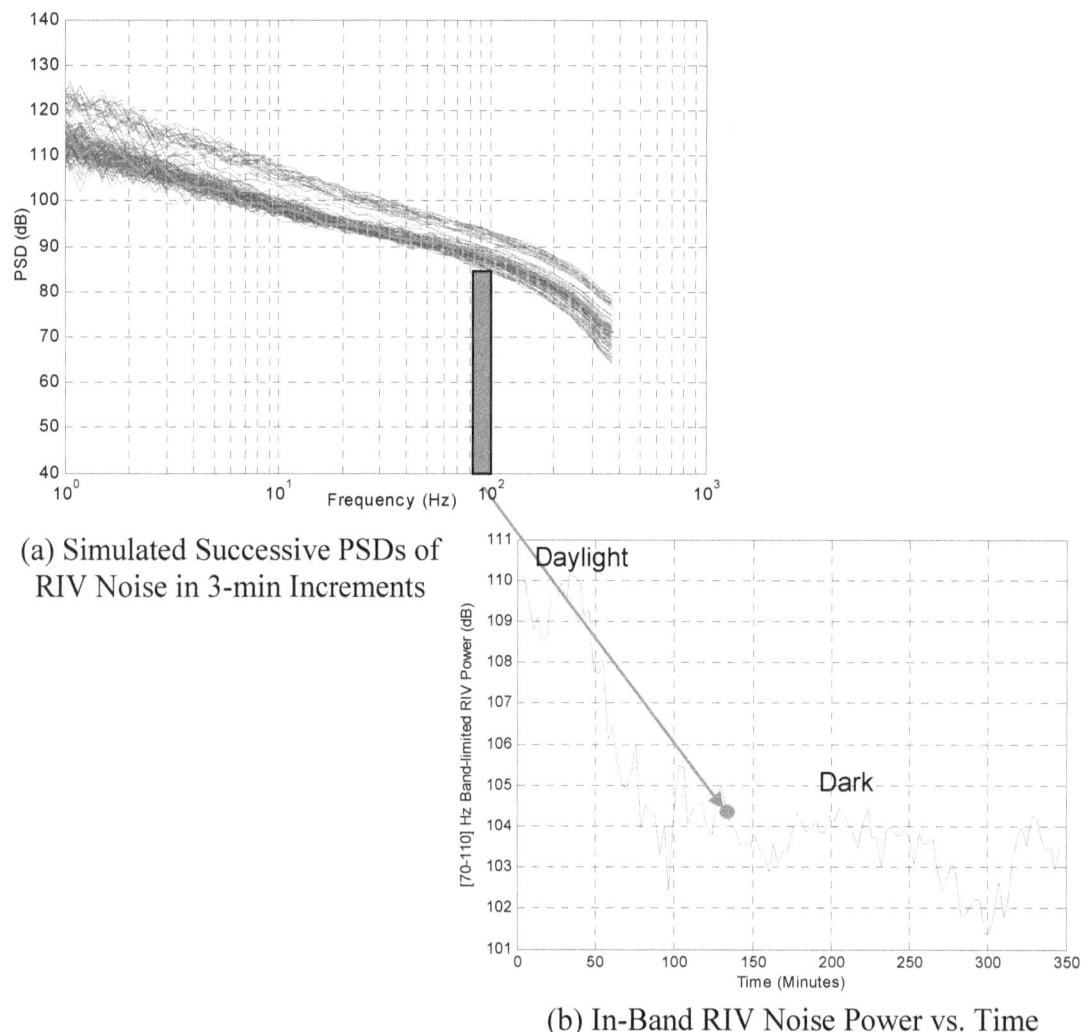

(a) Simulated Successive PSDs of RIV Noise in 3-min Increments

(b) In-Band RIV Noise Power vs. Time

Figure 4-9 Methodology for Simulating RIV Noise

It was assumed that there was no loss of coherence across each beam or from beam to beam, and that the Band-Pass Filter (BPF) for a beam's output had a 100-Hz width. Also, no other noise source except RIV noise was included. Thus, this simulation assumed a best-case scenario, and established upper bounds for the SNR expected to be obtained from processing real data.

4.3.4 Simulation Results

The three panels in Figure 4-10 depict the process by which the SNR associated with an aircraft wake and RIV noise were simulated. The first panel shows the predicted PSDs of the wake acoustic signals for skyward-looking arrays of one, four and eight beams, along with the RIV noise for the arrays. The second panel is the SNR at the output of the preprocessing 100 Hz BPF as a function of the filter's center frequency. The third panel is a time representation of the upper bound of simulated SNR variation over the approximately 6 hr for which turbulence data were available. It can be seen that eight beams were needed to achieve an SNR significantly above a 3 dB detection threshold for 5 hr of the 6 hr. These results represent an upper bound on SNR performance due to the idealized assumptions utilized.

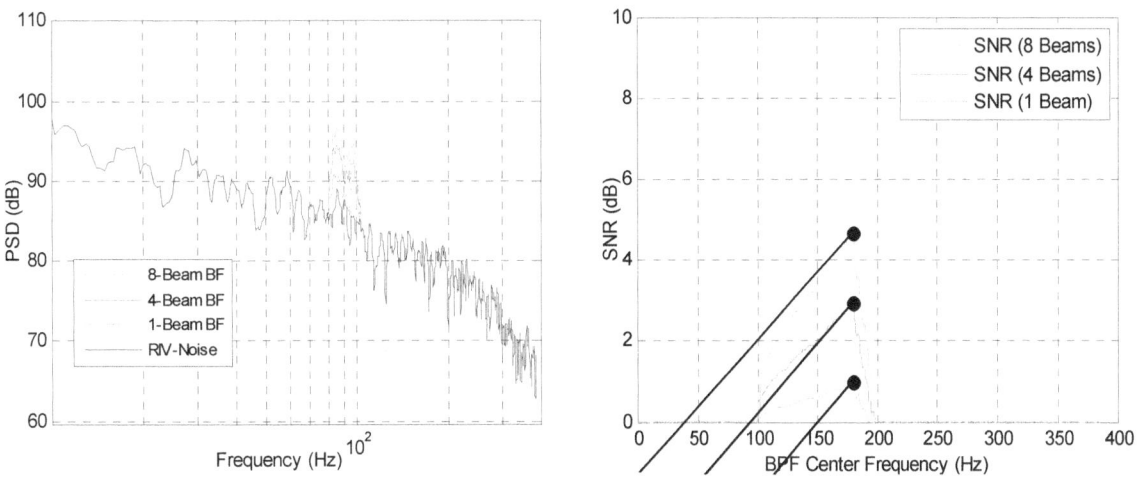

(a) Overlaid simulated PSD of a narrow band wake signal in the presence of RIV noise for 1, 4, and 8 beams

(b) Simulated SNR as a function of preprocessing BPF center frequency

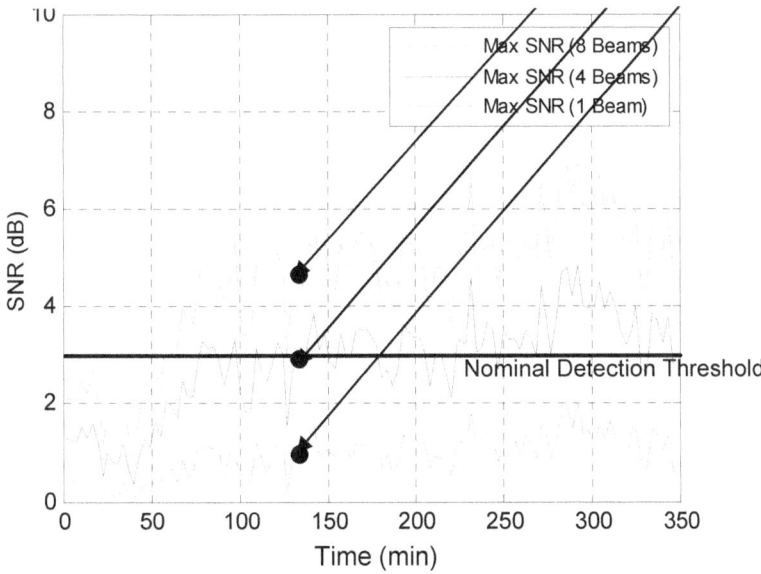

(c) Resulting maximum SNR vs. Time (when BPF center frequency matches that of the wake signal)

Figure 4-10 Methodology for Simulating Aircraft Wake and RIV Noise SNR

Figure 4-11 addresses the dependence of an eight-beam array's in-band SNR on wind speed and wind turbulence. Figure 4-11(a) shows that as the wind speed approaches 4.86 kt (2.5 m/sec), the simulated array SNR dips below the 3-dB nominal detection threshold. Since wind speed and EDR are highly correlated near the surface, an SNR dependence on EDR (a measure of atmospheric turbulence) is also expected. It can be seen in Figure 4-11(b) that when the EDR rises to about $3 \; 10^{-3} \; m^2/sec^3$, the array SNR upper bound is very likely to fall below the detection threshold. These plots suggest that multi-point wind speed or low-frequency turbulence measurements at a candidate site might be sufficient to predict how the sensor will perform relative to other sites and/or the times of the day and year that are optimum for sensor performance.

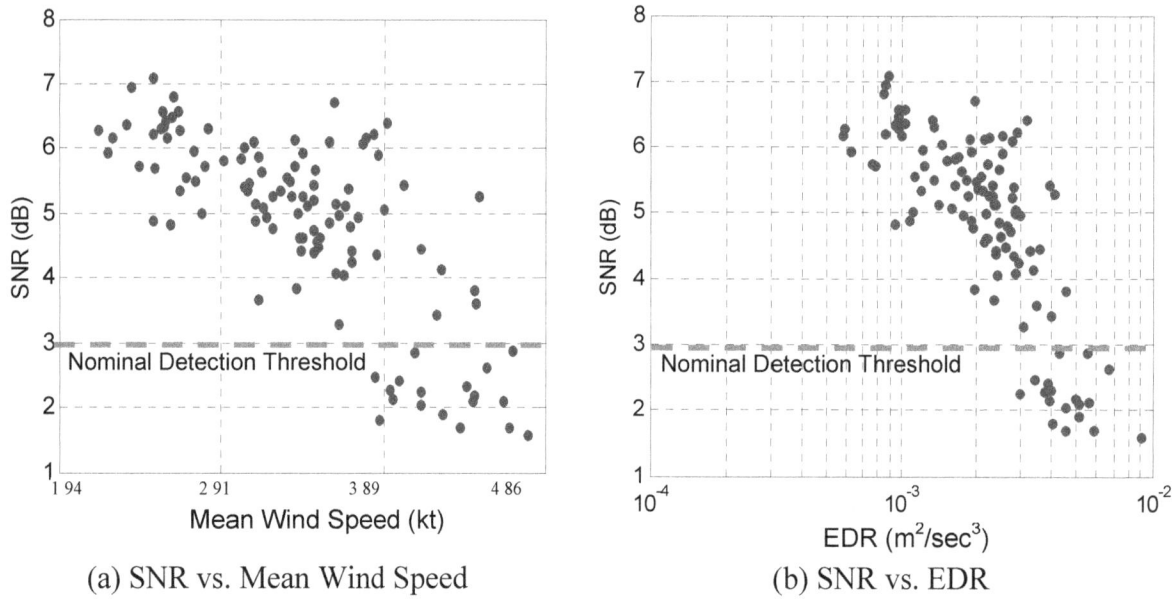

(a) SNR vs. Mean Wind Speed (b) SNR vs. EDR

Figure 4-11 Simulated Eight-Beam Array SNR vs. Wind Speed and EDR

Figure 4-12 is a plot of detection probability versus wind speed reported by FST for the DEN05 test (Ref. 3). It shows a decrease in detection for increasing wind speed, similar to that for SNR predicted by simulation. This is reasonable, since detection probability is related to SNR.

4.3.5 Interpretation of RIV Simulation Results

Simulations indicate a strong sensitivity of SOCRATES performance to ambient weather parameters. Over the course of less than 6 hr, under light wind conditions on a partly cloudy day, the data-driven simulation predicted an SNR change on the order of 8 dB. Extrapolating this trend suggests that SNR variations will be much larger for a hot summer day, high wind conditions, and/or in a strong rain. The simulation-based predictions are consistent with detection performance measured during the DEN05 test.

Given SOCRATES sensitivity to meteorological conditions, measurements of SOCRATES performance at one time, season, or location will not be reliable predictors of SOCRATES performance at another time/season/location. Instead, for each candidate new site, either direct measurements of SOCRATES performance, or estimates of its performance derived from meteorological sensors such as those deployed at DEN, must be obtained. Low-frequency

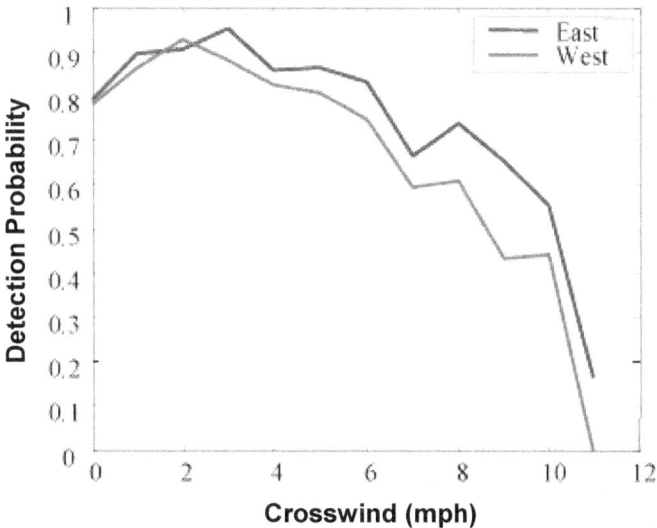

Figure 4-12 Measured DEN05 Detection Ratio vs. Crosswind (Ref. 3)

turbulence measures such as EDR proved to be correlated with the RIV noise floor. Thus, wind turbulence measurements could be used to predict SOCRATES performance in a new situation.

In the simulation employed herein, the wake was modeled as a fully coherent line source. A real wake might more closely resemble a line source that is not fully coherent, or with some variable coherence length; either would result in a lower SNR. Nevertheless, this simulation highlights the sensitivity of SOCRATES performance to meteorological effects.

4.4 Near-Field and Atmospheric Effects on Increased Array Size

This section analyzes the impact of increasing the length of a SOCRATES laser beam and/or increasing the number of beams in an array. Subsection 4.4.1 addresses the geometric effects of increasing the beam length when sensing a wake in the near-field; Subsection 4.4.2 addresses variations in the propagation path between the source and a laser beam for a constant wind and temperature profile; and Subsection 4.4.3 addresses the effects of turbulence along this propagation path.

Ideally, the SNR and spatial selectivity of a laser beam acting as an extended optical microphone will improve with beam length. Also, more beams, processed coherently, will ideally increase SNR. However, in practice, lengthening/widening an array will only improve performance when the following assumptions are true:

- **Assumption 1:** The acoustic source is in the beam's far field. That is, the incident waves are sufficiently planar that the actual beam length is less than the maximum effective beam length L_e — defined as that portion of a very long beam over which the received waves add non-destructively.

- **Assumption 2:** There is no significant distortion or bending of the acoustic wavefronts induced by wind and temperature changes/fluctuations between the source wake and sensing laser beam. When this is true, time/phase delays predicted based on the geometric distances between the acoustic source and various points along the array are almost equal to the actual delays and beamforming works well.

- **Assumption 3:** There is no loss of coherence across a single beam, or from one beam to another, due to turbulent temperature and velocity fields.

Assumptions 1 and 2 conflict to some extent. Satisfying the first requires that an acoustic source be distant enough for wavefront curvature to be insignificant over the length of a laser beam. Conversely, satisfying the second may require that the source be close enough to the laser beam (in order to avoid accumulated bending/distortion) that the plane wave assumption is not valid for the full length of the beam. Conditions to satisfy Assumption 3 are similar to those for Assumption 2, since loss of coherence increases with array aperture and distance to the source.

4.4.1 Maximum Effective Beam Length

Expression for Maximum Effective Beam Length — If Assumption 1 above is not satisfied, then the source is in the beam near field; in the case of a monopole source, the wavefronts then are spherical. Therefore, for a source broadside to the laser beam, the acoustic pressure will not be in-phase along the whole beam. A portion of the beam will contribute to the output signal, and a portion of the beam may degrade the output.

The maximum effective beam length for a point source, L_e, can be approximated by:

$$L_e = \sqrt{\lambda R}$$ **Equation 4-1**

In Equation 4-1, λ is the acoustic wavelength and R is the range between the source and the beam. This formula was derived based on the criterion that a maximum phase difference across the beam of $\varphi_c = \pi$ can be tolerated. This criterion was used because two coherent signals with a phase shift $\varphi_c = \pi$ will add destructively and will not contribute to the total output signal. For a monopole source, increasing the beam length beyond L_e will result in a reduction in SNR.

A plot of maximum effective beam length L_e versus range to the source R, for two frequencies of interest for wake acoustics, is shown in Figure 4-13. It is clear from this figure that the beam length for the DEN05 SOCRATES prototype significantly exceeded the maximum effective length for reasonable measurement scenarios. For example, a useful height for measuring wake

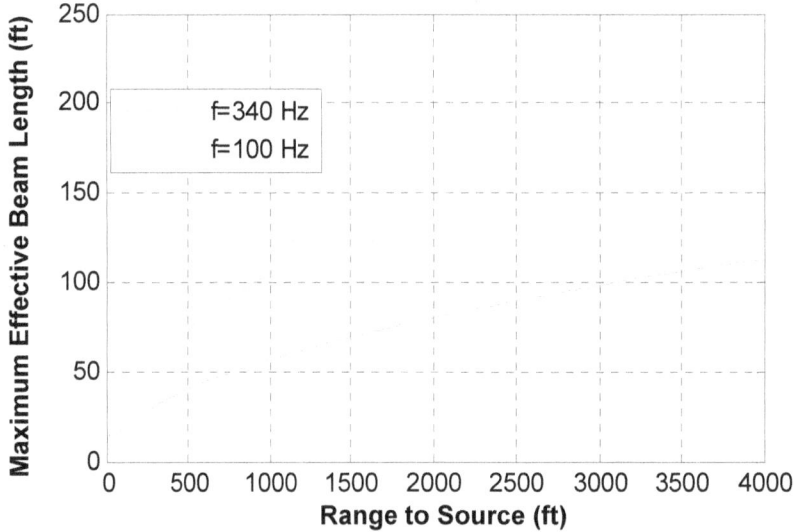

Figure 4-13 Maximum Effective Beam Length vs. Range to a Monopole Source
For a monopole source, if the beam length is increased beyond L_e no signal gain is achieved.

acoustics was 500 ft and a lower bound on the processing frequency was 100 Hz; for these values, the maximum effective beam length is 74 ft. Higher processing frequencies and lower altitude wakes would result in smaller maximum effective beam lengths. In contrast, the actual beam length was 328 ft.

Beam Response Sensitivity to Length (Point Source) — As the length L of a beam increases from $L=0$ to $L=L_e$, the response to an acoustic point source increases proportional to L^2 (20 dB/decade), as seen in Figure 4-14. In contrast, the uncorrelated RIV noise power at the beam output is proportional to L (10 dB/decade). As the beam length increases beyond L_e the beam response to the point source tends to flatten out, while the RIV noise coupled into the beam

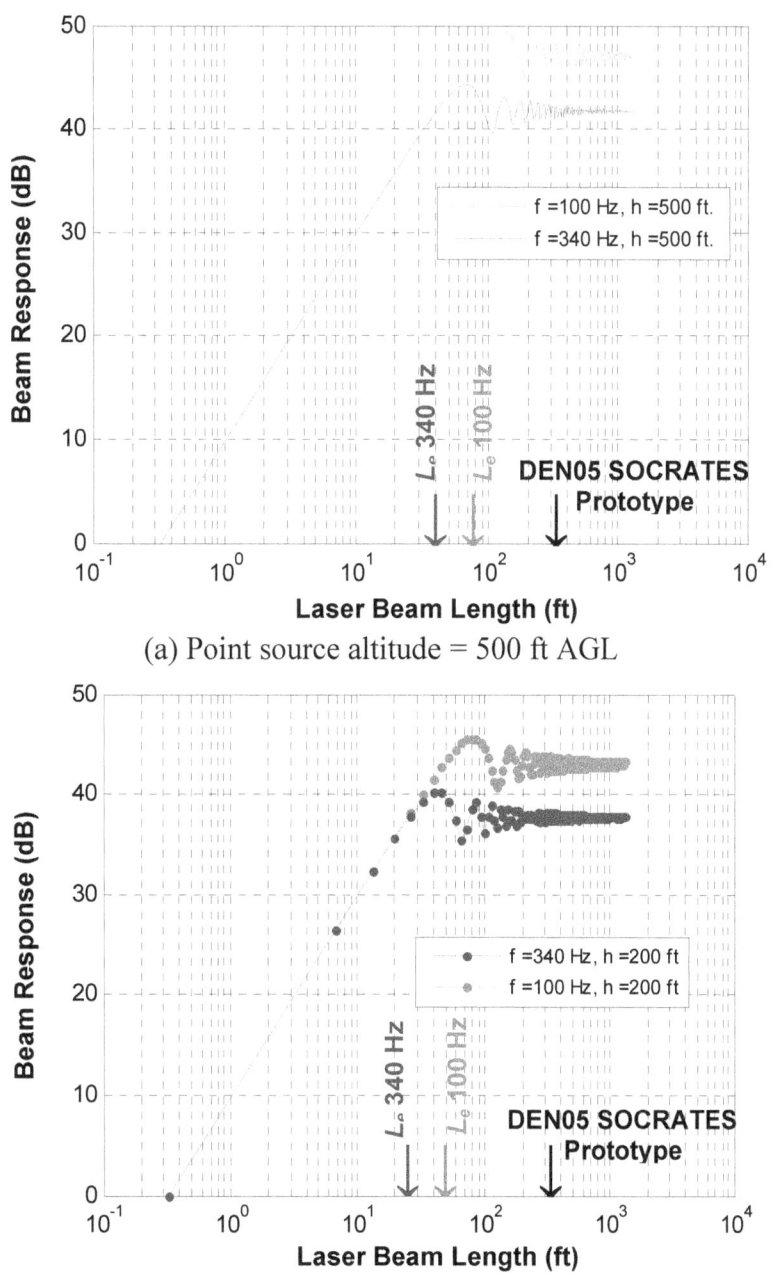

(a) Point source altitude = 500 ft AGL

(b) Point source altitude = 200 ft AGL

Figure 4-14 Beam Response to an Acoustic Point Source vs. Beam Length

continues to rise at the same rate. Therefore increasing the beam beyond a certain length will degrade SNR. The maximum effective beam length corresponds to a length where the beam response curve visually begins to depart from a straight line. Thus for the goal of selecting a beam length that optimizes SNR (for a particular source type, frequency, and altitude), the maximum effective beam length will generally be a reasonable choice.

Beam Response Sensitivity to Length (Line Source) — Simulations were also performed to investigate the sensitivity of the beam response to a 50 m (164 ft) coherent line source to changes in beam length (Figure 4-15). The simulation employed the same two source heights and frequencies that were considered previously. The line source was characterized by 150 point sources for the 100 Hz source frequency, and by 300 point sources for the 340 Hz source, to ensure that point-source spacing was less than half a wavelength. The laser beam was subdivided into 200 elements for the 100 Hz simulation, and 400 elements for the 340 Hz

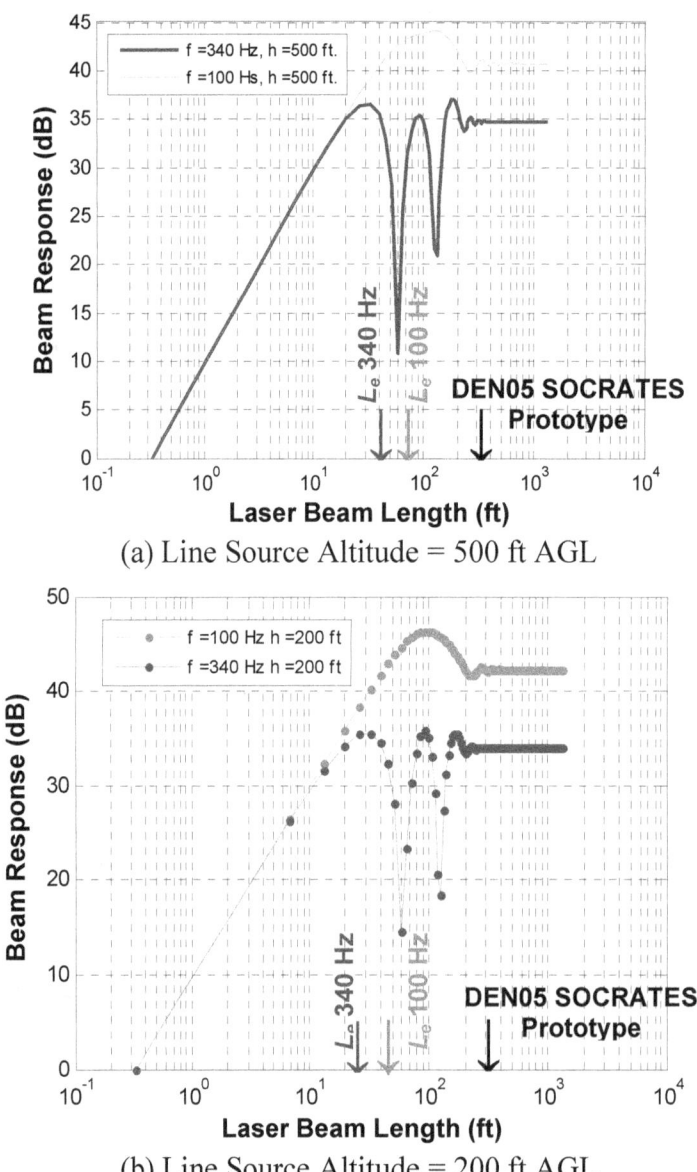

(a) Line Source Altitude = 500 ft AGL

(b) Line Source Altitude = 200 ft AGL

Figure 4-15 Beam Response to a 50-m Acoustic Line Source vs. Beam Length

simulation. It was verified that further increases in the spatial sampling resolution of either the line source or laser beam would not affect the results.

Based upon the length where the beam response curve visually begins to depart from a straight line, the maximum effective length for this line source is actually approximately one-half that for a point source (i.e., L_e given by Equation 4-1 and shown on the plots). Also, for the 340 Hz source, as the beam length increased beyond the maximum effective length, the response fluctuated dramatically (e.g., dropped by more than 25 dB for a wake at 500 ft) before stabilizing. In designing a SOCRATES-like system, it would be advisable to avoid this regime.

Beam Response Sensitivity to Line Source Orientation — Figure 4-16 shows how the beam response is predicted to vary with changes in wake orientation relative to the beam, for the line source and two heights/frequencies considered previously. (Simulations were performed using the same parameters as those used for Figure 4-15, except that number of sources forming the simulated line source was taken to be 600 for all frequency-height combinations, to ensure that the observed effects were not artifacts arising from a low spatial sampling rate of the line source.) In performing these simulations, the wake center-point was fixed above the beam's center-point. An angle of zero corresponded to the line source being parallel to the beam. The angle between the beam and source was changed by rotating the line source about its center-point in the horizontal plane (right part of Figure 4-16).

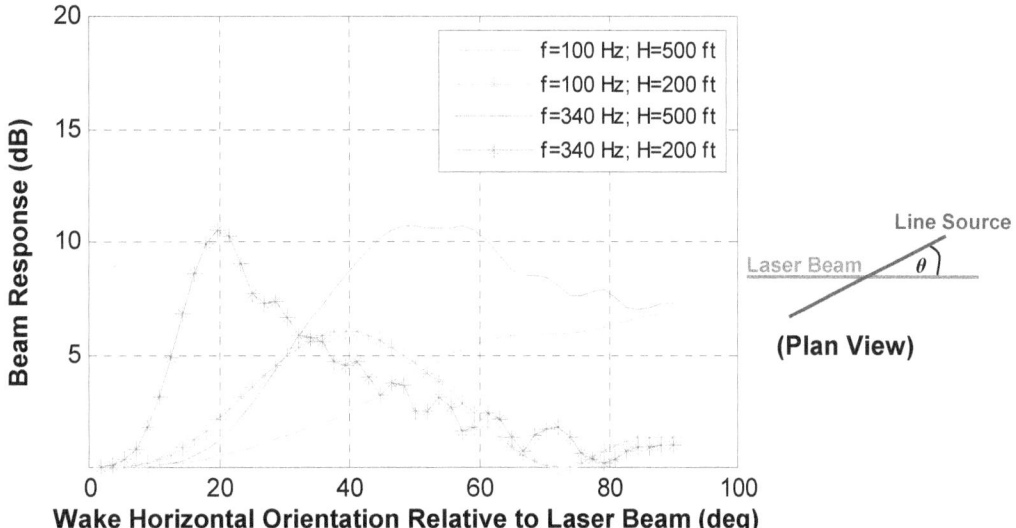

Figure 4-16 Laser Beam Response vs. Wake-Beam Orientation

The beam responses shown in Figure 4-16 are all normalized to the zero-angle case. While intuition might suggest that the maximum response would occur with the source parallel to the beam, this orientation in fact resulted in the minimum response. The "physical" explanation seems to be that a greater response occurs when a source is near the middle of the beam than when it is near the ends, so that rotating the line source from 0 deg increases the response. While there is no orientation that is best for all frequencies and heights, it appears that angles between 30 deg and 60 deg could increase the output from a wake acoustic signal by a few decibels.

Selection of Beam Length and Orientation — Figure 4-17 depicts the predicted SOCRATES beam response to point and line sources as a function of the source distance, for the same source frequencies that are considered above. As a newly formed wake moves closer to the beam and passes through the far field to near field transition region, received acoustic power can fluctuate

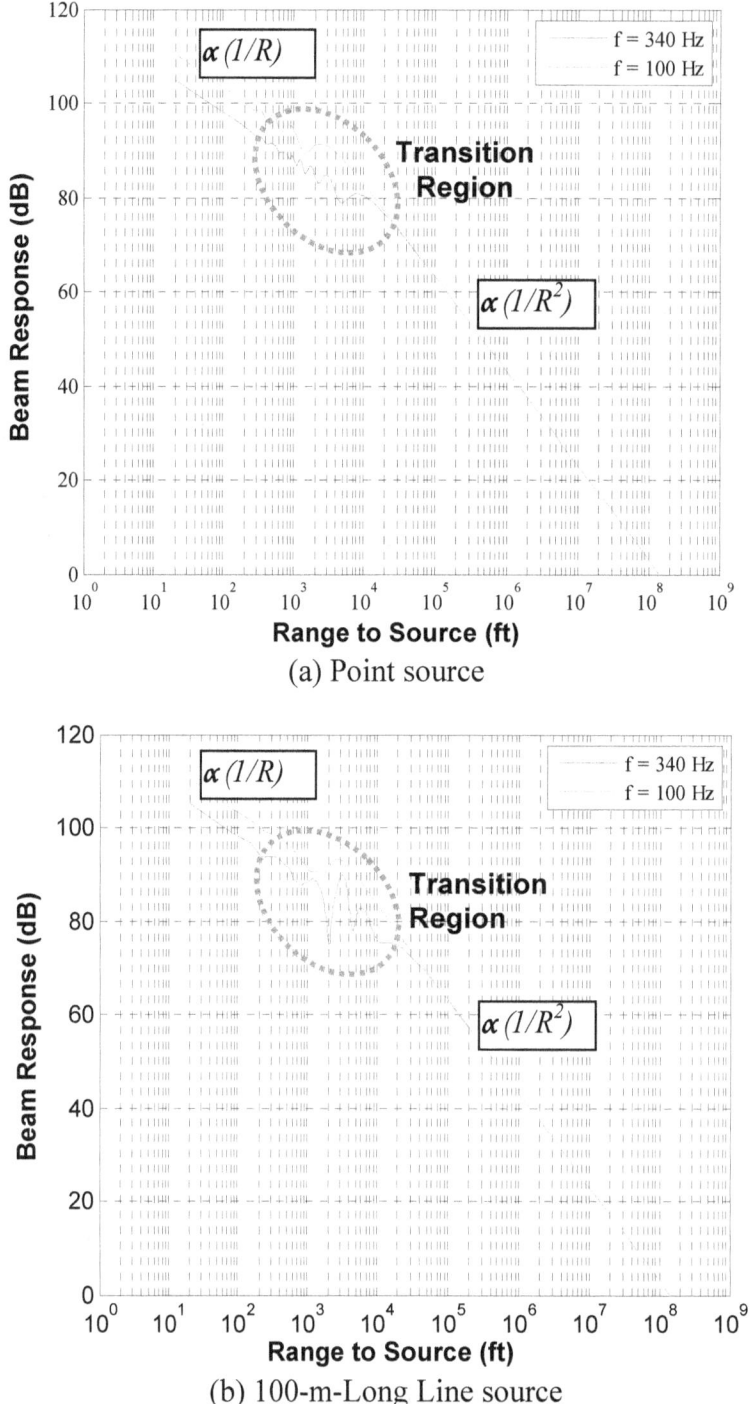

(a) Point source

(b) 100-m-Long Line source

Figure 4-17 Beam Response to Point and Line Sources vs. Source Range

by more than 15 dB. Also, these fluctuations can occur as Crow instabilities and non-uniform wind profiles cause a wake to change orientation relative to each sensing beam.

Due to the sensitivity of the beam response to wake range and orientation (Figure 4-16), such large variations in the detected acoustic power might cause a loss of detection. The severity of these events will depend on the tracking algorithm and the particular operational concept. Also, because the beam response variability happens without any changes in the source's emission characteristics, it will hamper the processor's ability to infer wake strength from received acoustic power.

In general, the beam length should be selected to avoid the response being anywhere in the transition region where it can fluctuate dramatically (e.g., Figure 4-15). Therefore the beam length should be slightly less than the potential effective length corresponding to the start of the transition region, or to a length that puts the response well beyond the transition region. The disadvantage of the latter is that as the beam length increases beyond the transition region, the noise increases without a corresponding gain in signal power. Also, additional simulations showed that the size of the transition region depends on the line source length, while the beam effective length prior to the transition region is little affected. This is further confirmation that for the case of a coherent wake source a beam length shorter than that employed in DEN05 should be used.

For a given line source length and acoustic frequency, the maximum effective length of the beam is a function of its distance from the source. As shown in Figure 4-17, for a fixed beam length, changing the source-to-beam distance can place the response in the near-field, transition region, or far-field. Therefore, selection of the beam length should consider the range of source-sensor distances, to ensure that the response does not enter the transition region as the wake descends below a predefined protection volume. For the DEN05 test, aircraft flew over the Skyward-Looking array at approximately 700 ft AGL. If, for example, wakes are to be measured in the height range of 200 to 700 ft, Figure 4-15(b) should be used to calculate the effective beam length for 200 ft (the lowest altitude of interest) which insures that for higher altitudes the response to a coherent wake will not enter the transition region regime. According to this same figure, this would be a length between 30 and 100 ft, depending on the upper bound of the processing frequency band utilized. Due to the large RIV noise floor masking the acoustic response up to 150 to 200 Hz, it is apparent that a beam length around 50 ft should be employed. However, the actual beam length, 328 ft, was considerably larger. Thus, in terms of SNR, the deployed SOCRATES system was sub-optimal and likely would have had better detection performance with shorter laser beams.

4.4.2 Aperture Size Limitation by Propagation Effects

Wind and temperature profiles will cause acoustic rays to bend and change speed as they propagate to different beams in an optical array. Therefore, as the source-beam and inter-beam separations increase, difference in propagation delays due to wind and temperature effects may be large enough to degrade beamforming performance. For the Skyward-Looking configuration listening to wakes at 200 to 700 ft altitudes, travel distances were too small for temperature and wind effects to be significant. However, for the Billboard configuration, distances to the wake were on the order of 3,500 ft, and propagation effects could have been important.

A simulation based on the SOCRATES Billboard configuration was carried out to investigate the maximum horizontal distance between two receivers (or two beam centers) for representative DEN05 wind and temperature profiles (Figure 4-18 and Figure 4-19, respectively). This simulation considered a source in the Billboard array region of focus — 3,200 ft (north), 400 ft (west), 1000 ft (vertical), relative to the Billboard array — and used ray-tracing to estimate the difference of time delays between two receivers relative to assumed delays.

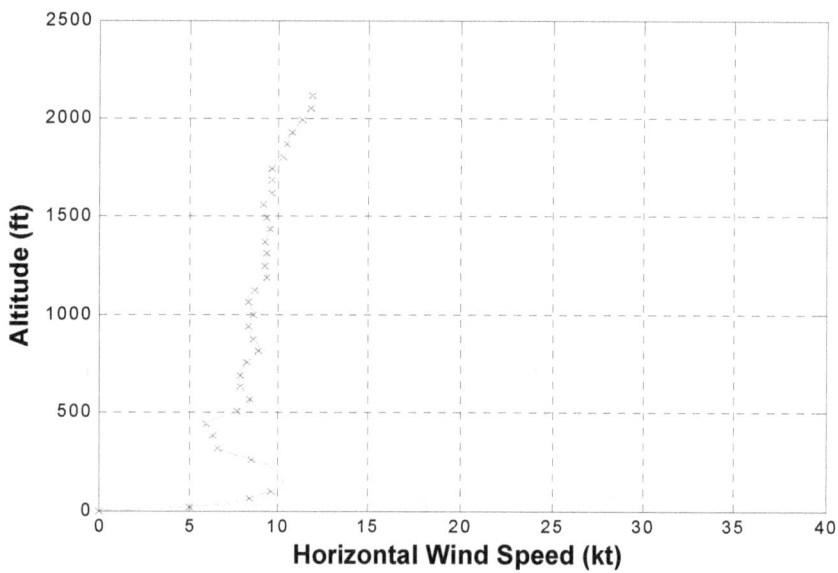

Figure 4-18 Representative Wind-Altitude Profile from DEN05 Test

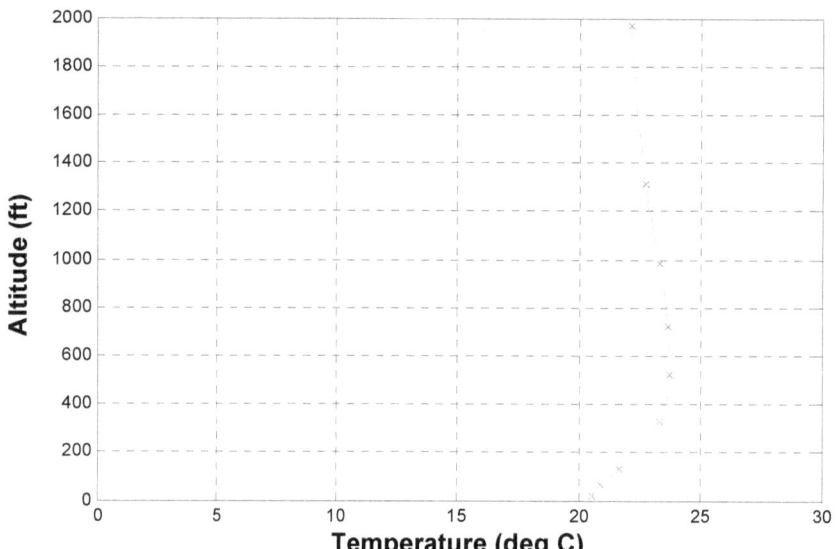

Figure 4-19 Representative Temperature-Altitude Profile from DEN05 Test

This simulation was conducted while varying the distance between the two receivers until a critical delay error was reached. The delay error corresponding to a phase shift of $2\pi/3$ was chosen, because for this error beamforming the two receivers will have no gain (0 dB). Table 4-3 summarizes this maximum horizontal distance for various frequencies, and can be used as a starting point in selecting the distance between centers of the vertical sub-arrays for the SOCRATES Billboard configuration. For example, for processing band of 200-400 Hz, and under these wind and temperature profiles, this table shows that not much benefit can be expected from increasing the horizontal separation between the centers of adjacent beams beyond approximately 150 ft. The aperture size of the DEN 05 Billboard array was significantly larger than that recommended based on this simulation, therefore it would not be expected to perform optimally if all sub-arrays were processed coherently.

Table 4-3 Billboard Array Recommended Maximum Sub-Array Separation vs. Frequency

Source Frequency (Hz)	Sub-Array Separation (ft)
100	436.4
200	217.5
300	145.3
400	108.9

4.4.3 Loss of Coherence Caused by Turbulent Boundary Layer

Wind and temperature turbulence encountered along the path from the sound source to different portions along the beam (or to different beam centers) will cause locations near one end to become uncorrelated with points near the other end. Loss of correlation can be another limiting factor on beam length. As in the case of ray-tracing effects, the maximum recommended distance between receivers (as well as maximum effective length L_e) will decrease with increasing distance to the source, making the Billboard configuration more susceptible to this effect.

5. Summaries and Conclusions

5.1 Summary: Wake Acoustic Phenomenology

The DEN03 campaign of wake acoustic measurements with a microphone array, combined with subsequent analyses, conclusively established the existence of acoustic emissions from aircraft wake vortices. The DEN03 test determined that the bulk of the acoustic energy usually was below 400 Hz, and was concentrated below 100 Hz for larger aircraft. Additionally, the predominant acoustic energy excess above the ambient noise background was for frequencies below 100 Hz. Source localization maps derived from the microphone array data showed that the predominant acoustic sources were near the vortices cores.

With the wake acoustic phenomenology established, a qualitative and quantitative evaluation of wake acoustics sensing, including SOCRATES, followed. The most meaningful evaluation of any sensor would involve a comparison to independently established requirements that have been developed from a validated operational concept. However, that was not possible in this situation. While several other wake sensors — e.g., Pulsed LIDAR, CW LIDAR, SODAR, and anemometer windline — have been employed as research instruments, their practicality as operational sensors for real-time use in controlling air traffic is considered questionable at this time. Reasons include performance limitations, lack of technical maturity, airport deployability issues, and complexity of operation. In the context of the FAA-NASA WTRMP, operational wake sensors are a Phase 3 activity and are scheduled to be addressed during 2010-2015.

Lacking operationally-based requirements, microphone and SOCRATES wake acoustic sensors are compared to each other and to other wake sensors in Section 5.2. In Section 5.3, issues associated with using SOCRATES in an operational environment are highlighted. Concluding statements are provided in Section 5.4.

5.2 Summary: Comparison of Acoustic Sensors to Other Wake Sensors

Since they rely on the same vortex phenomenology, the microphone array and SOCRATES wake detection sensors can/should logically be compared, and both can/should be compared with other wake sensors. In the DEN03 test, detection performances of the two sensors were similar: for a common set of 213 flybys, detection probability was 86% for the microphone array and 70% for the SOCRATES prototype. The SOCRATES contractor team has stated that, for all flybys of the SOCRATES prototypes, the achieved detection probabilities were: (a) 81% for DEN03, based on 882 flybys; and (b) 86% for DEN05, based on 1,987 flybys. Differences in detection performance were likely due to statistical fluctuations and different meteorological conditions.

A concern for both microphone array and SOCRATES sensors is that the wake acoustic signal energy has not been shown to be correlated to the strength of a wake. For example, in several instances, regional jets had stronger wake acoustic signals than larger Airbus A-318, A-319, A-320 and A-321 aircraft, and older aircraft were often more detectable than newer models. In addition to the lack of information about wake strength, this raises the possibility that wake acoustic detection may be less effective in the future than it is today, as new aircraft are introduced.

When contrasted with the wake tracks provided by a Pulsed LIDAR, the durations of microphone and SOCRATES-detected wakes were generally significantly shorter. Since a robust relationship between wake acoustic energy and aircraft hazard has not been identified, it is difficult to determine if termination of the wake tracks by both acoustic systems coincided with elimination, or at least a significant reduction, of the wake hazard.

The spatial resolution of the 252-microphone array was significantly better than that of either the DEN03 or DEN05 prototype SOCRATES sensors. The microphone array could resolve an aircraft's two vortices separately, and could distinguish changes in the state of the vortices. The SOCRATES sensors did not demonstrate either of these capabilities. Analysis indicated that a significant increase in the number of laser beams would be required for SOCRATES to be able to resolve two vortices — from the 16 deployed for DEN05 to between 40 and 320, depending upon the wake altitude and acoustic frequency of interest.

For operational use, the ability to distinguish two vortices during their descent would be an advantage, and possibly a requirement, for a system developed under Phase-3 of the WTRMP, as descent provides a mechanism for wake avoidance and vortices do not always sink as a pair. When considering research sensors, the ability to provide an image of the evolution of two line vortices is of scientific value. Previously, this capability was only available using aircraft mounted smokers or flybys of towers equipped with smokers.

To avoid low-frequency internal system noise, the DEN03 and DEN05 SOCRATES prototypes generally utilized the 270-360 Hz band for wake detection. However, for larger aircraft, wake acoustic signal energy was concentrated in the 0-100 Hz band that can be used by a microphone array. In the quiet Denver test environment this mismatch between signal and processing bands may not have had a significant impact on SOCRATES detection performance. In other measurement scenarios — e.g., louder background noise or greater distance between the wake and sensor — it is likely that a microphone array would have greater performance advantage.

The DEN03 and DEN05 tests were conducted under generally favorable conditions for wake acoustic sensors. The airport location provided a relatively pristine acoustic environment. Moreover, clear/dry low-wind VMC conditions prevailed almost throughout the test period. For these conditions, SOCRATES detection performance was found to be quite sensitive to cross-wind, solar heating, and temperature — e.g., dropped to 10% in 10 kt crosswind and to 60% in the mid-day sun. Performance of other wake sensors can also degrade with crosswind, but not to the same extent as the SOCRATES prototype.

An apparent advantage (not proven by testing) of acoustic sensors over LIDARs is the ability detect wakes in foggy or similar low ceiling/visibility conditions, or heavy rain. LIDARs are generally ineffective when ceiling/visibility is reduced significantly; the former primarily affects OGE vortex measurements and the latter affects IGE data collection. Any further field testing of the SOCRATES sensor should focus on IFR weather conditions, including the detection and tracking of vortices whose generation heights are above the weather ceiling.

5.3   Summary: Acoustics/SOCRATES Sensor Issues

The Denver test site was selected because it provided advantages for wake acoustic research that likely will not be present at many busy airports where a wake solution is needed. Issues associated with deploying wake acoustic/SOCRATES sensors at other airports are summarized here:

Acoustically Quiet Environment — The rural area surrounding DEN is among the quietest for major U.S. airports — e.g., much of the time a human observer could hear wakes immediately after their generation. Since this lack of ambient noise will not prevail near many busy urban/suburban airports, it is likely that acoustic sensor performance (i.e., wake detection probability and track duration) would be poorer, possibly significantly poorer, than for the DEN tests.

Available Real Estate Under Approach Path — In the Skyward-Looking configuration, the DEN05 SOCRATES sensor required two patches of land near the extended centerline of the runway being monitored, approximately 2.7 mi from the runway threshold. Each patch measured approximately 330 ft by 10 ft. Similarly, the 252-element microphone array required a 100 ft by 400 ft area under the glide slope. In urban settings, suitable patches at these specific or nearby locations may be difficult to acquire. Moreover, for runways having over-water approach paths, installation of an acoustic/SOCRATES wake detection sensor might not be practical or even possible.

Sensitivity to Non-Ideal Weather Conditions — Wake acoustic/SOCRATES testing to date has demonstrated detection probabilities in the 80% range during near-ideal weather conditions. However, during the DEN05 tests, SOCRATES detection performance degraded sharply in the presence of a crosswind or solar heating/elevated temperature. These sensitivities might be acceptable in a research context, but would have to be addressed for an operational sensor.

IGE Wake Measurements — Based on their measurement related siting requirements and processing technique, wake acoustic sensors, including SOCRATES, are better suited for detecting wakes that are generated in OGE conditions several hundred feet above the ground. There have been only a few efforts to detect and track vortices at lower altitudes using SOCRATES, and their interpretations were complicated by a limited understanding of both wake acoustic phenomenology and SOCRATES sensor characteristics. The interest in IGE wakes rests in the fact that it is this flight region where wake hazards are the greatest to following aircraft, and where wake turbulence separation is reinforced during IFR procedures. Any future testing and analysis of SOCRATES should address detection and tracking of IGE wakes.

Unattended Operation Not Demonstrated — During the entire DEN03 and DEN05 tests, the SOCRATES sensor was monitored by an engineer. While this is a reasonable operating mode for a sensor under development, it indicates the SOCRATES sensor is, in this aspect, less mature than some other wake sensors. For the past few years, Pulsed LIDAR, SODAR, and anemometer windlines have all operated unattended.

5.4 Conclusions: Acoustics/SOCRATES Suitability as Wake Sensors

At this time, it appears that the major contribution of the SOCRATES project is an increase in the foundational knowledge of wake acoustic phenomenology. This knowledge may lead to better understanding of wake dynamics, from which wake behavior can be better predicted.

When considered as potential research sensors, testing has shown that a phased microphone array is superior to a laser-based SOCRATES sensor. The NASA-DOT microphone array had slightly better detection performance and longer tracks. More importantly, it resolved both vortices, and showed their descent and evolutionary state changes.

When considered for real-time/operational use, there are important unresolved performance and "practical" issues for both types of wake acoustic sensors that would have to be considered:

- **Performance:** Inability to provide an indication of wake strength (wake acoustics), shortness of wake tracks (wake acoustics), inability to resolve individual wakes (SOCRATES only), and the overall shorter wake track relative to Pulsed LIDAR data, which is a serious drawback when it comes to certain operational concepts.
- **"Practical":** need for an acoustically quiet environment, demanding off-airport siting requirements, performance in non-ideal weather conditions, ability to detect wakes generated in ground effect, and inability to operate unattended.

Whether these are "show-stoppers" relative to inclusion in a specific system would depend upon that system's operational concept and derived requirements.

References

[1] Rudis, R.P., Wang, F.Y., and Daskalakis, A.C., <u>Status Report - SOCRATES Concept Exploration Effort</u>, DOT-VNTSC-RSPA-01-04, DOT/RITA/Volpe National Transportation Systems Center, Cambridge, MA, October 2001.

[2] Fiduccia, P.C., Bryant, W., and Lang, S., "FAA/NASA Wake Turbulence Research Program," *Journal of ATC*, January-March 2004, pp. 17-21.

[3] <u>Socrates Data Analysis Report: Denver 2005 Field Test</u>, Flight Safety Technologies, Inc., 2007.

[4] <u>Air Traffic Control</u>, Federal Aviation Administration, Order 7110.65R, February 16, 2006.

[5] Burnham, D.C., Clark, K.L., Hallock, J.N., Hannon, S.M., Jacobs, L.G., Rudis, R.P., Soares, M.A., and Wang, F.Y., <u>SFO Wake Turbulence Measurement System: Sensors and Data Descriptions</u>, DOT-VNTSC-FAA-07-03, October 2006, DOT/RITA/Volpe National Transportation Systems Center, Cambridge, MA.

[6] Cotton, W., and Williams, R., "Project SOCRATES: A New Sensor Technology for Enhancement of Aviation Safety and Capacity," The 2002 FAA Airport Technology Conference, Atlantic City, NJ, May 5-8, 2002.

[7] Garodz, L.J., and Clawson, K.L., "Vortex Wake Characteristics of B757-200 and B767-200 Aircraft Using the Tower Fly-By Technique, Volume 1" NOAA Technical Memorandum ERL ARL-199, January 1993.

[8] Bedard, A.J., Jr., and Georges, T.M., "Atmospheric Infrasound," *Physics Today*, Issue 53, March 2000, pp. 32-37.

[9] Burnham, D., Gorstein, M., Hallock, J.N., Kodis, R., Sullivan, T., and McWilliams, I.G., <u>Aircraft Wake Vortex Sensing Systems</u>, DOT-TSC-FAA-72-13, June 1971, DOT Transportation Systems Center, Cambridge, MA.

[10] Bedard, A.J., Jr., "Low-Frequency Atmospheric Acoustic Energy Associated with Vortices Produced by Thunderstorms," *Monthly Weather Review*, Volume 133, January 2005, pp. 241-263.

[11] Fine, N., Rees, F., and Von Winkle, W., "Acoustic Measurements in an Airport Environment and the Identification of the Sound Radiated by Aircraft Wake Vortices," *Journal of the Acoustical Society of America*, vol. 103, no. 5, pt. 2, May 1998, p. 2803.

[12] Rubin, W. L., "The Generation and Detection of Sound Emitted by Aircraft Vortices in Ground Effect," *Journal of Atmospheric and Oceanic Technology*, Volume 22, May 2005, pp. 543-554.

[13] Michel, U., and Böhning, P., "Investigation of Aircraft Wake Vortices With Phased Microphone Arrays," AIAA Paper No. 2002-2501, 8th AIAA/CEAS Aeroacoustics Conference, Breckenridge, CO, June 2002.

[14] Braukus, M., and Barnstoff, K., "Scientists Listen in on Sounds of Aircraft Wake Signature," NASA Press Release 03-271, August 19, 2003.
http://www.nasa.gov/lb/home/hqnews/2003/aug/HQ_03271_ACFT_WAKE.html

[15] Wang, F.Y., Wassaf, H., Dougherty, R.P., Clark, K., Gulsrud, A., Fenichel, N., and Bryant, W.H., "Passive Wake Acoustics Measurements at Denver International Airport," *4th NASA Integrated CNS Conference and Workshop*, Fairfax, VA, April 2004.

[16] Dougherty, R.P., Wang, F.Y., Booth, E.R., Watts, M.E., Fenichel, N., and D'Errico, R.E., "Aircraft Wake Vortex Measurements at Denver International Airport," AIAA Paper No. 2004-2880, 10th AIAA/CEAS Aeroacoustics Conference, Manchester, UK, May 2004.

[17] Wassaf, H., Gulsrud, A., Wang, F. Y., Chalson, M., and Egan, J., "Wake Acoustics Data Analysis at USDOT Volpe Center – Status Update," *Proceedings of the NASA Wake Acoustics Test Program Review and Mini-Workshop, Denver International Airport*, August 18-19, 2004.

[18] Wassaf, H. S., Ibe, O.C., Dougherty, R.P., and Zhang, Y., "Acoustical Spectral Analysis of a Wake Vortex Cross-Section Using Microphone-Arrays," *The Journal of the Acoustical Society of America*, Vol. 117, Issue 4, April 2005, p. 2546.

[19] Wang, F.Y., Wassaf, H.S., Gulsrud, A., Delisi, D.P., and Rudis, R.P., "Acoustic Imaging of Aircraft Wake Vortex Dynamics," AIAA Paper No. 2005-4849, 23rd AIAA Applied Aerodynamics Conference, Toronto, Canada, June 2005.

[20] Booth, E.R., Jr., and Humphreys, W.M., Jr., "Tracking and Characterization of Aircraft Wakes Using Acoustic and LIDAR Measurements," AIAA Paper No. 2005-2964, 11th AIAA/CEAS Aeroacoustics Conference, Monterey, California, May 2005.

[21] Humphreys, W.M., Jr., and Booth, E.R., Jr., "Wake Acoustic Microphone Array Analysis," Proceedings of the NASA Wake Acoustics Test Program Review and Mini-Workshop, Denver International Airport, August 18-19, 2004.

[22] Dougherty, R.P., "Extensions of DAMAS and Benefits and Limitations of Deconvolution in Beamforming," AIAA Paper No. 2005-2961, 11th AIAA/CEAS Aeroacoustics Conference, Monterey, California, May 2005.

[23] Wassaf, H.S., Zhang, Y., and Ibe, O.C., "Wake Acoustic Analysis and Image Decomposition via Beamforming of Microphone Signal Projection on Wavelet Subspaces," 12th AIAA/CEAS Aeroacoustics Conference, Cambridge, Massachusetts, May 2006.

[24] Böhning, P., "Akustische Lokalisierung von Wirbelschleppen," Doctoral Dissertation (*in German*), Technical University of Berlin, April 2006.

[25] Fine, N.E., and Kring, D.C., "Opto-Acoustic Tracking of Aircraft Wake Vortices," AIAA Paper No. 2005-2965, 11th AIAA/CEAS Aeroacoustics Conference, Monterey, California, May 2005.

[26] Zheng, Z. C., Li, W., Wang, F.Y., and Wassaf, H., "Influence of Vortex Core on Wake Vortex Sound Emission," AIAA Paper No. 2006-2538, 12th AIAA/CEAS Aeroacoustics Conference, Cambridge, Massachusetts, May 2006.

[27] Lilley, G.M., "A Quest for Quiet Commercial Passenger Transport Aircraft for Take-off and Landing," AIAA Paper No. 2004-2922, 10th AIAA/CEAS Aeroacoustics Conference, Manchester, UK, May 2004.

[28] Gerz, T., Holzäpfel, F., Bryant, W., Köpp, F., Frech, M., Tafferner, A., and Winckelmans, G., "Research Towards a Wake-Vortex Advisory System for Optimal Aircraft Spacing," C. R., physique, Vol.6, pp. 501-523, 2005.

[29] Loucel, R.E., and Crouch, J. D., "Flight-Simulation Study of Airplane Encounters with Perturbed Trailing Vortices," *Journal of Aircraft*, Vol. 42, No.4, July-August 2005, pp. 924-931.

[30] Bush, J., Cekorich, A., and Kirkendall, C. "Multi-Channel Interferometric Demodulator," http://www.optiphase.com/techlib.htm

[31] Cekorich, A., "Demodulator for Interferometric Sensors," http://www.optiphase.com/techlib.htm

[32] Gladstone, J.H., and Dale, T.P., Philosophical Transactions of the Royal Society, volume 153, (1863), pp 317-343.

www.ingramcontent.com/pod-product-compliance
Lightning Source LLC
Chambersburg PA
CBHW081834170526
45167CB00007B/2797